UNLEASHING THE POWER OF THE OODA LOOP IN CYBERSECURITY

A New Frontier in Cybersecurity: Applying the OODA Loop and Implicit Guidance and Control

Bob Maley

Master Your Cyber Defense: Observe, Orient, Decide, Act!

Printed in the United States of America.

For more information or to book an event, contact:
Bob@c-ooda.com
http://www.c-ooda.com

ISBN – Print Edition 979-8-9881956-1-0
ISBN – Kindle 979-8-9881956-0-3
First Edition: July 2023
Editor – Marya Roddis, CTPRP, S.U.N. Resource Development

Foreword

As the author of this book, I am both thrilled and honored to share my journey with the OODA Loop and its application in the field of cybersecurity. Over the past decade, I have been an avid student of the OODA Loop, exploring its intricacies and discovering its vast potential for enhancing decision-making processes in diverse domains.

This book aims to serve as a comprehensive guide to the OODA Loop, exploring its history, fundamental principles, and applications in cybersecurity. By sharing my knowledge and experience, I aim to empower cybersecurity professionals and organizations alike to harness the power of the OODA Loop to enhance their decision-making processes, adapt to the rapidly changing cyber threat landscape, and, ultimately, safeguard their digital assets and infrastructure.

The sources used in this book have been instrumental in shaping my understanding of the OODA Loop and its relevance to cybersecurity. These sources have provided valuable insights and served as the foundation for developing my expertise in applying the OODA Loop principles to the ever-evolving world of cyber threats and defenses.

In crafting this book, I have employed the power of ChatGPT, an advanced AI language model developed by OpenAI, to analyze and summarize the various sources that have inspired my work. ChatGPT has allowed me to create an accessible and comprehensive guide to the OODA Loop and its application in cybersecurity, offering readers a unique perspective on this crucial subject.

However, I must emphasize that the insights and analysis provided by ChatGPT differ from the deep understanding I have gained through my years of experience as a student of the OODA Loop. I have applied my knowledge and the sources I have studied to discern the accuracy and value of ChatGPT analysis and ensure that the information presented in this book is correct and relevant.

I invite you to join me on this journey to explore the fascinating world of the OODA Loop and its potential for transforming how we approach cybersecurity. May the insights and strategies you discover within these pages serve as a catalyst for your own growth and success in cyber security.

Table of Contents

About the Author

"Success is not final, failure is not fatal: it is the courage to continue that counts." - Winston Churchill

Bob Maley, Inventor, CISO, Author, Futurist, and OODA Loop fanatic is currently the Chief Security Officer at Black Kite, a technology company that specializes in cybersecurity intelligence gathering and analysis used in assessing potential impact of cyber events at third parties.

Bob has been involved in security for most of his career, initially in physical security as a law enforcement officer. He has acquired a broad range of experience and expertise across all areas of security, including third-party security and risk assessment; IT architecture, design, policy development, deployment, incident response and investigation; and enterprise solution deployments in areas including intrusion detection, data protection, compliance, and incident reporting and response.

Before assuming the Chief Security Officer role at Black Kite, Bob was the head of PayPal's Global Third-Party Security & Inspections team, developing the program from the ground up into a state-of-the-art risk management program.

In a previous role as chief information security officer for the Commonwealth of Pennsylvania, he led the Pennsylvania Information Security Architecture program to win the 2007 award for outstanding achievement in information technology by the National Association of State Chief Information Officers (NASCIO).

Bob has been named a CSO of the Year finalist for the SC Magazine Awards and was nominated as the Information Security Executive of the Year, North America. Additionally, his team was a Best Security Team finalist in the SC Magazine Awards. Bob's certifications include CRISC, CTPRP, OpenFAIR™ and CCSK. His expertise has been quoted in numerous articles, including Forbes, Payments.com, StateTech Magazine, SC Magazine, Wall Street

Journal, Washington Post, Dark Reading, and many others. His publications include whitepapers, the IEEE Journal, and his first book, What Every Engineer Should Know About Cybersecurity and Digital Forensics, which was released in December 2022.

Introduction

"Knowing is not enough, we must apply. Willing is not enough, we must do." - Johann Wolfgang von Goethe

In today's hyper-connected world, organizations and individuals face an ever-evolving landscape of cyber threats. From phishing emails to ransomware, data breaches to nation-state cyber-espionage, the stakes have never been higher when it comes to protecting our digital assets and infrastructure. In my more than 20 years of experience in the cybersecurity field, I have witnessed firsthand the rapid development of cybersecurity challenges and innovative approaches to tackling these threats. I have seen the rise of cybercrime and cyber warfare, and their impact on global economies, national security, and individual privacy. My passion for understanding and addressing the complexities of the cybersecurity domain has led me to explore a groundbreaking decision-making framework known as the OODA Loop, developed by Colonel John Boyd, which has the potential to transform the way we approach cybersecurity and incident response.

As the digital world continues to grow and become more complex, so does the challenge of safeguarding it. To address these challenges, we need to adopt a forward-thinking, proactive approach that goes beyond traditional cybersecurity measures. We need to embrace frameworks and methodologies that enable us to adapt and respond rapidly to ever-changing threats.

Colonel John Boyd's OODA Loop

The OODA Loop is one such framework. In essence, the OODA Loop—Observe, Orient, Decide, and Act—is a continuous cycle of perceiving information, analyzing, and synthesizing it, making decisions based on the analysis, and then acting. This decision-making framework was conceived by Colonel John Boyd, an American Air Force pilot and military strategist.

The OODA Loop consists of four Phases: Observe, Orient, Decide, and Act. Each step plays a critical role in the overall decision-making process, and the continuous cycling through these Phases allows for ongoing adaptation and improvement. The strategies underlying the OODA Loop, such as Communication and Collaboration and Continuous Improvement, resonate across all of the four Phases of the OODA Loop process.

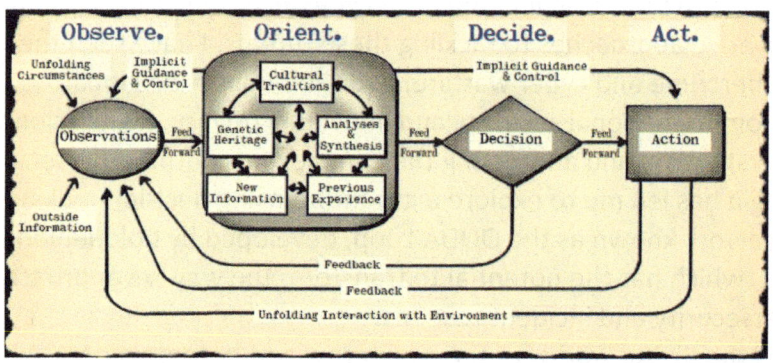

Most people think OODA is a simple 4 stage process loop, it is far more than that.

The OODA Loop has been successfully applied in various fields—from military combat to business strategy. Implicit Guidance and Control (IGC), another concept introduced by Boyd, further refines the decision-making process by emphasizing the importance of intuition, experience, and rapid response in situations characterized by uncertainty and ambiguity.

In this book, I explore the relevance and potential applications of the OODA Loop and Implicit Guidance and Control in cybersecurity. We will delve into the intricacies of each Phase of the OODA Loop, discuss its inherent adaptability and iterative nature, and investigate how integrating IGC into the decision-making process can significantly enhance the quality and speed of our responses to cyber threats. Furthermore, we will examine the practical implications of implementing the OODA Loop and IGC in cybersecurity programs, focusing on fostering collaboration, agility, and resilience in the face of rapidly evolving challenges.

As we navigate this landscape, we must recognize that

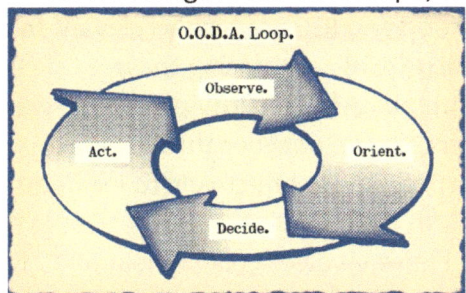

cybersecurity is not merely a technical issue; rather it is a multifaceted challenge that demands a holistic approach. By adopting the OODA Loop and Implicit Guidance and Control as guiding principles in our cybersecurity strategies, we can empower organizations and individuals to make better decisions, respond more effectively to incidents, and ultimately build a more secure digital landscape.

The Cyber Landscape

The cybersecurity landscape has undergone a dramatic transformation over the past two decades. In the early days of the internet, cybersecurity threats primarily consisted of isolated incidents of hacking, often perpetrated by individuals seeking to exploit vulnerabilities in computer systems for personal gain or simply for the thrill of it. Fast forward to today, and the scope and scale of cyber threats have grown exponentially. Cybercriminals have become more sophisticated, leveraging advanced techniques and tools to infiltrate systems, steal sensitive data, and disrupt critical infrastructure. Nation-states have also entered the fray, engaging in cyber espionage and cyber warfare to gain a strategic advantage over adversaries.

This ever-evolving quality of cyber threats calls for a dynamic, adaptive approach to cybersecurity. Traditional security measures, such as firewalls, antivirus software, and intrusion detection systems—while still essential components of a comprehensive cybersecurity program—are no longer sufficient to protect against various advanced threats that organizations and individuals face daily. We must adopt those measures and frameworks that enable us to think and act more strategically and effectively in the face of uncertainty, ambiguity, and rapid change.

This is where the OODA Loop and Implicit Guidance and Control come into play. Originally developed for military applications, the principles underlying the OODA Loop and IGC are highly relevant to the challenges we face in cybersecurity. By embracing these concepts and integrating them into our cybersecurity programs, we can better understand the dynamic nature of cyber threats, make more informed decisions, and respond more effectively to incidents as they occur.

In cybersecurity, the **Observe** Phase involves collecting and monitoring data from various sources, such as network logs, threat intelligence feeds, and security alerts, to gain a real-time understanding of the current threat landscape. This Phase requires a robust and comprehensive approach to data collection, as the quality of the information gathered will directly impact the subsequent steps of the OODA Loop.

In the **Orient** Phase analysis, synthesis, and understanding come into play. Here, the information collected during the Observe Phase is processed and contextualized, considering factors such as cultural traditions, genetic heritage, new information, previous experiences, and analytical insights. This Phase is critical, enabling us to make sense of the data we have gathered and identify patterns, trends, and potential threats that may require action.

The Orient Phase also involves considering how our own biases, assumptions, and mental models may influence our interpretation of the data and adjust our perspective accordingly to ensure a more accurate and complete understanding.

Next comes the **Decide** Phase, in which we formulate response strategies based on our understanding of the situation. This Phase requires a delicate balance between speed and accuracy, as making decisions too quickly may lead to suboptimal outcomes, while conversely taking too long may result in missed opportunities to mitigate or prevent damage from cyber threats. By integrating Implicit Guidance and Control principles into the decision-making process, we can leverage our intuition, experience, and knowledge to make more rapid and effective decisions, even amid uncertainty and limited information.

Finally, the **Act** Phase involves executing the chosen response to address identified threats. Acting may include deploying patches to fix vulnerabilities, blocking malicious IP addresses, or initiating incident response procedures to contain and remediate a breach. The effectiveness of the Act Phase depends heavily on the quality of the decisions made during the Decide Phase and the speed at which these decisions can be implemented.

This iterative, feedback-driven approach is particularly well-suited to the rapidly changing and unpredictable nature of the cybersecurity landscape. As we cycle through the OODA Loop, we continuously learn from our experiences and interactions with the environment, allowing us to adapt and improve our planning, decision-making, and response capabilities.

Implementing the OODA Loop and Implicit Guidance and Control in a cybersecurity program requires a concerted effort to:

- build a skilled and agile Cybersecurity Incident Response Team (CIRT),
- develop a comprehensive Incident Response Plan,
- leverage advanced Security Information and Event Management (SIEM) systems,
- cultivate a security-aware organizational culture,
- conduct regular cybersecurity assessments and threat intelligence analyses, and
- encourage collaboration and information sharing among cybersecurity stakeholders.

By adopting these practices, organizations can enhance their ability to detect, prevent, and respond to cyber threats, ultimately improving their overall security posture and resilience.

Investigating the interplay between human factors and technology in the context of the OODA Loop and IGC can provide useful insight into how we can enhance our ability to make better decisions and respond more effectively to cyber threats. As we look to the future, ongoing research and development in areas such as Artificial Intelligence (AI) and Machine Learning (ML) will undoubtedly play a significant role in shaping the cybersecurity landscape and how we approach threat detection and mitigation. These advanced technologies hold great promise for enhancing our ability to analyze vast amounts of data, identify patterns and anomalies, and make rapid, informed decisions in the face of ever-evolving cyber threats. Integrating AI and ML into the OODA Loop and IGC frameworks may further strengthen their effectiveness in addressing the unique challenges of the cybersecurity domain.

Another critical area for future research and exploration is the impact of human factors on cybersecurity decision-making. As the OODA Loop and IGC emphasize the importance of intuition, experience, and rapid response, understanding how human cognitive processes, biases, and emotions influence decision-making and incident, response is essential for optimizing the application of these frameworks in cybersecurity programs. Additionally, it is important to evaluate the efficacy of the OODA Loop and IGC implementation in various cybersecurity scenarios and industry sectors. This may involve conducting case studies — several of which are provided later in the book—and simulations or real-world experiments to assess the impact of these frameworks on the speed, quality, and outcomes of cybersecurity decision-making and incident response. Such research can help refine and adapt the OODA Loop and IGC principles to address better the specific needs and challenges of different industries and organizations.

In conclusion, the OODA Loop and Implicit Guidance and Control offer a transformative approach to enhancing cybersecurity decision-making and incident response in a world characterized by rapidly changing threats and increasing digital interconnectedness. By embracing these frameworks and integrating them into our cybersecurity strategies, we can empower organizations and individuals to make better decisions, respond more effectively to incidents, and build a more secure digital landscape. The critical need for continuous innovation, research, and adaptation in the face of evolving cyber threats underscores the importance of advancing our understanding and application of the OODA Loop and IGC, as well as fostering collaboration and knowledge-sharing among cybersecurity stakeholders. Together, we can work towards a safer and more resilient digital world for all.

CHAPTER 1: The Observation Phase

"The only true source of knowledge is experience." - Albert Einstein

A Comprehensive Approach to Data Collection in Cybersecurity

The rapidly evolving landscape of cyber threats presents a formidable challenge for organizations and individuals seeking to protect their digital assets and infrastructure. To effectively detect, prevent, and respond to these threats, it is essential to adopt a forward-thinking, proactive approach that goes beyond traditional cybersecurity measures. One such approach is the OODA Loop, a proven decision-making framework developed by Colonel John Boyd, which emphasizes the importance of continuous adaptation and rapid response in the face of uncertainty and change. The OODA Loop consists of four Phases: Observe, Orient, Decide, and Act. In this section, we will focus on the first Phase—Observe—and explore its implications for cybersecurity with an emphasis on developing a comprehensive approach to data collection and monitoring to understand the ever-changing threat landscape better.

The Importance of the Observe Phase in Cybersecurity

In the context of cybersecurity, the Observe Phase involves collecting and monitoring data from various sources to gain a real-time understanding of the current threat landscape. This Phase is critical, as the quality of the information gathered will directly impact the subsequent Phases of the OODA Loop, ultimately affecting the effectiveness of our decisions and response actions. Moreover, the ability to observe and analyze data in real time allows us to identify emerging threats and vulnerabilities, enabling us to take proactive steps to mitigate or prevent potential attacks before they occur.

Developing a Comprehensive Approach to Data Collection

To ensure the effectiveness of the Observe Phase in a cybersecurity program, it is essential to develop a comprehensive approach to data collection that encompasses a wide range of information sources and types. This includes both internal and external data, as well as structured and unstructured data, which together can provide insights into key factors relating to the threat landscape and inform our understanding of potential risks and vulnerabilities.

Internal Data Sources

Internal data sources are those that originate within an organization's own systems and networks. These can include, but are not limited to:

1. Network logs: Network logs contain records of events and transactions that occur within an organization's network, such as user login attempts, file downloads, and network connections. Analyzing network logs can help identify suspicious activity or patterns that may indicate a potential security breach or vulnerability.

2. Security alerts: Security alerts are generated by various security tools and systems, such as intrusion detection systems (IDS), firewalls, and antivirus software. These alerts can provide valuable information about potential threats and vulnerabilities, enabling security teams to take appropriate action to mitigate or prevent attacks.

3. System and application logs: System and application logs contain records of events and transactions that occur within an organization's computer systems and software applications. This can include information about software updates, configuration changes, and user activities. Analyzing system and application logs can help identify potential security risks and vulnerabilities, as well as provide insights into user behavior and system performance.

4. Incident reports: Incident reports are created by security teams or other employees when they detect or respond to a

security event or breach. These reports can provide valuable information about the nature of the incident, the actions taken to address it, and any lessons learned that can be applied to improve security in the future.

External Data Sources

External data sources are those that originate outside an organization's systems and networks. These can include, but are not limited to:

1. Threat intelligence feeds: Threat intelligence feeds are streams of data that provide information about known threats and vulnerabilities, often sourced from external security research organizations, government agencies, and industry groups. These feeds can help security teams stay informed about the latest threats and vulnerabilities, allowing them to take proactive steps to protect their systems and networks. As CSO at Black Kite, I would be remiss in not advising you to look at what they do in this space. https://blackkite.com.

2. Security blogs, forums, and mailing lists: security professionals and researchers share information, insights, and experiences related to cybersecurity threats, vulnerabilities, and best practices through these online resources. By participating in these communities and staying informed about the latest developments and trends, security teams can enhance their knowledge and understanding of the threat landscape and better prepare for potential attacks.

3. Social media, websites, and news sources: These platforms can also provide valuable information about emerging cybersecurity threats and incidents. By monitoring these sources, security teams can stay informed about the latest attacks, breaches, and vulnerabilities, as well as gain insights into potential future threats and trends.

4. Industry and government cybersecurity reports: These reports often contain in-depth analyses and recommendations, which can provide useful insight into the latest threats, vulnerabilities, and best practices and can

help security teams better understand the threat landscape and improve their own security posture.

5. Information sharing and analysis centers (ISACs): ISACs are member-driven organizations that facilitate the sharing of threat intelligence and best practices among organizations within specific industries or sectors. By participating in an ISAC, organizations can gain access to valuable threat intelligence and insights specific to their industry or sector, enabling them to better protect their systems and networks. As a former member of the Multi-State ISAC (MS-ISAC), I recommend them as a great resource for anyone in the public sector (**https://www.cisecurity.org/ms-isac**).

Structured and Unstructured Data

In addition to considering a wide range of internal and external data sources, it is important to recognize the value of both structured and unstructured data in the Observe Phase of the OODA Loop. This distinction of structured vs. non structured refers to different forms of internal and external data. Structured data refers to information organized in a specific format, such as spreadsheets, databases, or log files. Unstructured data refers to information that does not have a specific format or structure, such as text documents, emails, or social media posts.

Both structured and unstructured data can provide valuable insights into the threat landscape and inform our understanding of potential risks and vulnerabilities. For example, structured data from network logs or security alerts can help identify patterns of suspicious activity that may indicate a potential security breach, while unstructured data from social media or news sources can provide insights into emerging threats and trends.

To effectively analyze and process both structured and unstructured data, organizations may need to leverage advanced data analytics tools and techniques, such as data mining, machine learning, and natural language processing. These technologies can help security teams more efficiently and effectively extract valuable insights from large volumes of data, enabling them to make better-informed decisions and respond more quickly to potential threats and vulnerabilities.

Importance of Observe Phase

The Observe Phase of the OODA Loop is a critical component of an effective cybersecurity program, as it enables organizations to gather and analyze data from a wide range of sources to better understand the ever-changing threat landscape. By developing a comprehensive approach to data collection that encompasses both internal and external data sources, as well as structured and unstructured data, organizations can enhance their ability to detect, prevent, and respond to cyber threats.

In addition to collecting and monitoring data, it is important for organizations to continuously refine and improve their observation capabilities by investing in advanced analytics tools and technologies, as well as fostering a culture of collaboration and information-sharing among security teams and stakeholders. By doing so, organizations can ensure that their cybersecurity programs remain agile and adaptive in the face of rapidly evolving threats and challenges, ultimately contributing to a more secure and resilient digital landscape.

Building a Robust Monitoring and Data Collection Infrastructure

To support the comprehensive approach to data collection and analysis discussed earlier, organizations must invest in building a robust monitoring and data collection infrastructure. This infrastructure should be capable of ingesting, processing, and analyzing data from a wide variety of sources, while also ensuring the scalability, reliability, and security of the systems involved. The following components and best practices are essential for building an effective monitoring and data collection infrastructure:

1. Security Information and Event Management (SIEM) Systems: SIEM systems are a critical component of a robust monitoring infrastructure, as they enable organizations to collect, analyze, and correlate security events and log data from multiple sources in real-time. By implementing a SIEM system, organizations can gain a holistic view of their security posture, identify patterns of suspicious activity, and quickly respond to potential threats and vulnerabilities.

2. Intrusion Detection and Prevention Systems (IDPS): IDPS solutions are designed to monitor network traffic and identify potential threats or malicious activity. By deploying an IDPS solution, organizations can gain increased visibility into their network environment, allowing them to detect and respond to potential security incidents more effectively.

3. Endpoint Detection and Response (EDR) Solutions: EDR solutions provide organizations with increased visibility and control over their endpoint devices, such as laptops, desktops, and mobile devices. By implementing an EDR solution, organizations can monitor and analyze endpoint activity, detect potential threats, and quickly respond to security incidents.

4. Data Collection and Aggregation Tools: To efficiently process and analyze data from multiple sources, organizations should invest in data collection and aggregation tools. These tools can help gather and normalize data from various sources, ensuring that data can be easily analyzed and correlated by SIEM systems, analytics tools, and security teams.

5. Data Analytics and Machine Learning Solutions: As mentioned earlier, advanced data analytics and machine learning solutions can play a significant role in enhancing an organization's ability to analyze and process large volumes of structured and unstructured data. By investing in these technologies, organizations can extract valuable insights from their data, enabling them to make better-informed decisions and respond more quickly to potential threats and vulnerabilities.

6. Continuous Monitoring and Improvement: Building an effective monitoring and data collection infrastructure is not a one-time project; rather this is an ongoing process of continuous improvement. Organizations should regularly evaluate the effectiveness of their monitoring infrastructure, identify areas for improvement, and invest in new technologies and best practices to ensure that their

systems remain capable of detecting and responding to the latest threats and challenges.

Fostering a Culture of Collaboration and Information-Sharing

In addition to building a robust monitoring and data collection infrastructure, organizations should also foster a culture of collaboration and information-sharing among security teams, stakeholders, and industry partners. By sharing threat intelligence, best practices, and lessons learned from security incidents, organizations can enhance their collective understanding of the threat landscape and improve their ability to detect and respond to emerging threats and vulnerabilities.

Several initiatives and platforms can facilitate this collaboration and information-sharing, including:

1. Information Sharing and Analysis Centers (ISACs): As mentioned earlier, ISACs are member-driven organizations that facilitate the sharing of threat intelligence and best practices among organizations within specific industries or sectors. By participating in an ISAC, organizations can gain timely access to valuable threat intelligence and insights that are specific to their industry or sector, enabling them to better protect their systems and networks.

2. Cybersecurity Frameworks and Standards: Adopting industry-standard cybersecurity frameworks and guidelines, such as the NIST Cybersecurity Framework or the ISO/IEC 27001 standard, can help organizations establish a common language and set of best practices for addressing cybersecurity risks and challenges. By aligning their cybersecurity programs with these frameworks and standards, organizations can enhance their ability to collaborate and share information with industry partners and stakeholders.

3. Public-Private Partnerships: Collaborating with government agencies and public-sector organizations can broaden understanding of and resources for addressing cybersecurity

challenges. Public-private partnerships can facilitate information-sharing, joint research initiatives, and the development of shared threat intelligence platforms, enabling organizations to better understand and respond to the evolving threat landscape.

4. Internal Collaboration and Cross-Functional Teams: Fostering a culture of collaboration within an organization is also essential for effective cybersecurity. This may involve creating cross-functional teams that include members from different departments, such as IT, security, legal, and human resources. By working together, these teams can share expertise, insights, and perspectives, ultimately leading to more effective decision-making and incident response.

5. Security Awareness and Training Programs: Ensuring that all employees within an organization are aware of the latest cybersecurity threats, best practices, and their role in maintaining security is crucial. By investing in security awareness and training programs, organizations can foster a culture of vigilance and information-sharing among their employees, enhancing their ability to detect, mitigate, and respond to potential security incidents.

The Observe Phase of the OODA Loop is a critical component of an effective cybersecurity program, as it enables organizations to gather and analyze data from a wide range of sources to better understand the ever-changing threat landscape. By developing a comprehensive approach to data collection, building a robust monitoring and data collection infrastructure, and fostering a culture of collaboration and information-sharing, organizations can enhance their ability to detect, prevent, and respond to cyber threats.

Ultimately, the success of the Observe Phase in a cybersecurity program depends on the organization's commitment to continuous improvement, investment in advanced technologies and best practices, and collaboration with industry partners and stakeholders. By embracing these principles, organizations can ensure that their cybersecurity programs remain agile and adaptive in the face of rapidly evolving threats and challenges, contributing to a more secure and resilient digital landscape.

Chapter Sources

1. **Boyd, J. R. (1987). A Discourse on Winning and Losing. Unpublished lecture notes, Air University Library, Maxwell AFB, Alabama.**
2. National Institute of Standards and Technology (NIST). (2018). Framework for Improving Critical Infrastructure Cybersecurity, Version 1.1. Retrieved from https://nvlpubs.nist.gov/nistpubs/CSWP/NIST.CSWP.041620 18.pdf
3. International Organization for Standardization (ISO) and International Electrotechnical Commission (IEC). (2013). ISO/IEC 27001:2013 - Information technology - Security techniques - Information security management systems - Requirements. Retrieved from https://www.iso.org/standard/54534.html
4. **Stiennon, R. (2015). There Will Be Cyberwar: How The Move To Network-Centric Warfighting Has Set The Phase For Cyberwar. IT-Harvest Press.**
5. **Zetter, K. (2014). Countdown to Zero Day: Stuxnet and the Launch of the World's First Digital Weapon. Crown.**
6. Lockheed Martin Corporation. (n.d.). The Cyber Kill Chain. Retrieved from https://www.lockheedmartin.com/en-us/capabilities/cyber/cyber-kill-chain.html
7. SANS Institute. (n.d.). Critical Security Controls. Retrieved from https://www.sans.org/critical-security-controls
8. Caltagirone, S., Pendergast, A., & Betz, C. (2013). The Diamond Model of Intrusion Analysis. Retrieved from https://apps.dtic.mil/sti/pdfs/ADA586960.pdf
9. Mandiant. (n.d.). M-Trends 2022: Annual Threat Report. Retrieved from **https://www.mandiant.com/sites/default/files/2022-04/M-Trends%202022%20Executive%20Summary.pdf**

10. Ponemon Institute. (n.d.). Cost of a Data Breach 2022 Report. Retrieved from https://www.ibm.com/security/data-breach
11. Information Sharing and Analysis Centers (ISACs). (n.d.). Retrieved from https://www.nationalisacs.org/member-isacs
12. Black Kite: The Cost of a Data Breach, A New perspective. 2022 https://blackkite.com/wp-content/uploads/2022/08/2022_CostofaDataBreachReport_BlackKite.pdf

CHAPTER 2: The Orientation Phase

"The only true wisdom is in knowing you know nothing." - Socrates

A Deep Dive into the Importance of Context and Analysis in Cybersecurity

The Orientation Phase of the OODA Loop focuses on making sense of the vast amount of data collected during the Observation Phase. The Orientation Phase helps organizations contextualize the data, allowing security teams to analyze and synthesize the information to form a comprehensive understanding of the cyber threat landscape. In doing so, the Orientation Phase enables organizations to identify vulnerabilities, predict and assess potential threats, and develop effective strategies to mitigate risks.

This in-depth exploration of the Orientation Phase will discuss the importance of context and analysis in cybersecurity, the role of cultural traditions, genetic heritage, new information, previous experiences, and the processes of analysis and synthesis in shaping an organization's understanding of and response to cyber threats.

The Importance of Context and Analysis in Cybersecurity

The Orientation Phase in the OODA Loop emphasizes the importance of context and analysis in making sense of the vast amount of data collected during the Observation Phase. Context and analysis help organizations identify patterns and trends, understand the motives and tactics of adversaries, and assess the potential impact of threats on their operations and assets.

1. Identifying Patterns and Trends: The analysis of data collected during the Observation Phase can reveal patterns and trends in cyber threats and vulnerabilities. By examining these patterns and trends, organizations can gain insights into the tactics, techniques, and procedures (TTPs) employed by adversaries and develop effective strategies to counter these threats.

2. Understanding Adversaries' Motives and Tactics: The context provided by the Orientation Phase helps organizations understand the motives and tactics of their adversaries. This understanding is crucial in developing targeted and effective strategies to defend against cyber-attacks. For instance, understanding the motivation of a nation-state attacker might involve analyzing geopolitical tensions, while understanding the tactics of a cybercriminal group could involve examining their past activities and preferred targets.

3. Assessing the Potential Impact of Threats: During the Orientation Phase, organizations can assess the potential impact of cyber threats on their operations and assets. This assessment allows organizations to prioritize their efforts, allocate resources efficiently, and develop tailored strategies to address the most significant risks.

The Role of Cultural Traditions and Genetic Heritage

Cultural traditions and genetic heritage play a significant role in shaping an organization's approach to cybersecurity, as they influence the values, beliefs, and priorities that guide decision-making processes. Understanding the influence of cultural traditions and genetic heritage on an organization's cybersecurity strategy can help security teams better understand the context in which they operate and develop more effective strategies to address emerging threats.

1. Cultural Traditions: An organization's cultural traditions can impact its approach to cybersecurity in several ways. For instance, the level of risk tolerance within a company, as well as its approach to innovation and collaboration, can be influenced by cultural factors. By recognizing and understanding these cultural influences, security teams can better tailor their strategies to align with the organization's values and priorities.

2. Genetic Heritage: Genetic heritage refers to the inherited characteristics, experiences, and knowledge passed down

through generations. In the context of cybersecurity, an organization's genetic heritage can include its history of dealing with cyber threats, the legacy technologies and systems it employs, and the institutional knowledge of its security teams. Understanding an organization's genetic heritage can help security teams identify potential vulnerabilities and areas for improvement, as well as leverage the organization's unique strengths and experiences in developing their cybersecurity strategies.

The Role of New Information and Previous Experiences

New information and previous experiences play a guiding role in shaping an organization's understanding of the cyber threat landscape and informing their decision-making processes during the Orientation Phase.

1. New Information: As the cyber threat landscape evolves rapidly, staying informed of the latest threats, vulnerabilities, and best practices is crucial for organizations. New information, such as threat intelligence reports, research findings, and industry news, can illuminate emerging trends and tactics employed by adversaries. By incorporating this new information into analysis and decision-making processes, organizations can enhance the ability to anticipate and respond to evolving threats and challenges.

2. Previous Experiences: An organization's previous experiences with cyber threats and incidents can provide valuable insights into the effectiveness of their cybersecurity strategies and their ability to respond to and recover from security incidents. By examining past incidents and learning from their successes and failures, organizations can identify areas for improvement, refine strategies, and better prepare for future threats.

The Processes of Analysis and Synthesis in the Orientation Phase

The Orientation Phase involves two key processes: analysis and synthesis. These processes enable organizations to make sense of the vast amount of data collected during the Observation Phase and develop a comprehensive understanding of the cyber threat landscape.

1. Analysis: The analysis process involves breaking down the collected data into smaller, more manageable pieces to identify patterns, trends, and relationships. This process can involve various techniques, such as data mining, statistical analysis, and machine learning algorithms, to uncover valuable insights hidden within the data. The analysis process also involves assessing the credibility and reliability of the data sources, as well as identifying any potential biases or inaccuracies that may impact the organization's understanding of the threat landscape.

2. Synthesis: The synthesis process involves combining the results of the analysis with the organization's cultural traditions, genetic heritage, new information, and previous experiences to form a comprehensive understanding of the cyber threat landscape. The synthesis process helps organizations contextualize the data, identify potential threats and vulnerabilities, and develop targeted and effective strategies to address these risks. Synthesis may involve the use of scenario planning, threat modeling, and risk assessments to evaluate the potential impact of threats on the organization's operations and assets.

Challenges and Best Practices for the Orientation Phase

The Orientation Phase of the OODA Loop presents several challenges for organizations, including wading through the sheer volume and complexity of the data to be analyzed, the rapidly evolving nature of the cyber threat landscape, and the need to balance the competing priorities and demands of various stakeholders. To address these challenges, organizations should consider the following best practices:

1. Implementing Advanced Analytics and Machine Learning Technologies: As mentioned earlier, advanced analytics and machine learning technologies can significantly enhance an organization's ability to process and analyze large volumes of structured and unstructured data. By investing in these technologies, organizations can extract valuable insights from their data more efficiently, enabling them to make better-informed decisions and respond more quickly to emerging threats and vulnerabilities.

2. Fostering a Culture of Collaboration and Information-Sharing: As discussed earlier, collaboration and information-sharing are essential for effective cybersecurity. By fostering a culture of collaboration within their organization and partnering with industry peers, government agencies, and other stakeholders, organizations can enhance their collective understanding of the threat landscape and improve their ability to respond to emerging threats and challenges.

3. Encouraging Continuous Learning and Improvement: The rapidly evolving nature of the cyber threat landscape requires organizations to be agile and adaptive in their approach to cybersecurity. Encouraging continuous learning and improvement, both at the individual and organizational levels, can help security teams stay informed of the latest threats, vulnerabilities, and best practices, and ensure that their strategies remain effective and up to date.

The Orientation Phase of the OODA Loop is a critical component of an effective cybersecurity program, as it enables organizations to make sense of the vast amount of data collected during the Observation Phase and develop a comprehensive understanding of the cyber threat landscape. By focusing on the importance of context and analysis, the role of cultural traditions, genetic heritage, new information, and previous experiences, and the processes of analysis and synthesis, organizations can enhance their ability to anticipate and respond to emerging threats and challenges.

Chapter Sources

1. Endsley, M. R. (1995). Toward a theory of situation awareness in dynamic systems. Human Factors, 37(1), 32-64.
2. Tavani, H. T. (2011). Ethics and Technology: Controversies, Questions, and Strategies for Ethical Computing. John Wiley & Sons.
3. Zhang, Y., & Wen, J. (2016). The IoT electric business model: Using blockchain technology for the internet of things. Peer-to-Peer Networking and Applications, 10(4), 983-994.
4. Goodman, M., & Kirk, J. (2017). Cyberspace and Cybersecurity: Foundations of Digital Forensics. CRC Press.
5. Chismon, D., & Ruks, M. (2015). Threat Intelligence: Collecting, Analysing, Evaluating. MWR InfoSecurity.
6. Chen, H., Chiang, R. H., & Storey, V. C. (2012). Business intelligence and analytics: From big data to big impact. MIS Quarterly, 36(4), 1165-1188.
7. Axelrod, C. W. (2015). Outsourcing Information Security. Artech House.
8. Casey, E. (2011). Digital evidence and computer crime: Forensic science, computers, and the internet. Academic Press.
9. DeFranco, J & Maley,B (2022). What Every Engineer Should Know About Cyber Security and Digital Forensics, 2nd Edition

CHAPTER 3: The Decide Phase

"In any moment of decision, the best thing you can do is the right thing, the next best thing is the wrong thing, and the worst thing you can do is nothing." - Theodore Roosevelt

Navigating Complexity, Uncertainty, and Time-Sensitive Challenges with the OODA Loop

The Decision Phase of the OODA Loop is a critical juncture that sets the course for action, and in the context of cybersecurity, it plays a pivotal role in determining how an organization responds to various cyber threats. The Decision Phase is where organizations select the most appropriate course of action based on the insights and understanding developed during the Orient Phase. A thorough examination of the Decision Phase, its key components, challenges, and strategies to enhance decision-making in cybersecurity is essential for organizations to maximize the effectiveness of the OODA Loop. The analysis of this Phase also demonstrates why the OODA Loop is so valuable in parsing through the layered complexity of the process.

The Nature of Decision-Making in Cybersecurity

The Decision Phase in cybersecurity involves making choices that can have significant implications for an organization's digital assets, reputation, and overall security posture. Decisions in cybersecurity are often characterized by:

- High stakes: The consequences of poor decision-making in cybersecurity can be severe, leading to financial losses, operational disruptions, and reputational damage.
- Complexity: Cybersecurity decisions often involve multiple variables, interdependencies, and trade-offs that must be carefully considered and balanced.

- Uncertainty: The rapidly evolving cyber threat landscape and the often-stealthy nature of cyberattacks create a high degree of uncertainty in decision-making.
- Time pressure: Cybersecurity decisions often need to be made quickly to prevent or mitigate ongoing attacks and minimize potential damage.

These factors make decision-making in cybersecurity a complex and challenging process that demands expertise, adaptability, and a robust decision-making framework like the OODA Loop.

Key Components of the Decision Phase

To make effective decisions in cybersecurity, organizations need to consider several key components:

- Objectives: Clearly defined objectives help guide decision-making by providing a framework for evaluating potential actions and their alignment with organizational goals.
- Constraints: Decision-makers need to consider the constraints that may limit their options, such as budgetary limitations, legal and regulatory requirements, and technological capabilities.
- Alternatives: Identifying a range of potential actions enables organizations to evaluate the relative merits and drawbacks of each option before selecting the most appropriate course of action.
- Evaluation: A systematic evaluation process that considers the potential benefits, risks, and trade-offs of each alternative is essential for informed decision-making.
- Decision criteria: Establishing decision criteria, such as effectiveness, cost, time, and alignment with organizational objectives, helps organizations prioritize and rank alternatives to make better informed choices.
- Feedback and adaptation: Continuous feedback and adaptation are essential for refining decision-making processes and improving the organization's ability to respond effectively to emerging cyber threats.

Challenges in Decision-Making for Cybersecurity

Organizations face several challenges in making effective decisions in the context of cybersecurity:

- Information overload: The sheer volume of data and information that organizations must process and analyze can lead to information overload, making it difficult to identify and prioritize relevant information for decision-making.
- Cognitive biases: Human decision-makers are prone to cognitive biases, such as confirmation bias, anchoring, and availability bias, which can distort their perception of risks and lead to suboptimal decisions.
- Groupthink: In group decision-making settings, the phenomenon of groupthink can lead to conformity and a suppression of dissenting opinions, which can hinder the development of innovative and effective solutions.
- Limited resources: Organizations often have limited resources, such as time, budget, and personnel, which can constrain their ability to explore and implement optimal security measures.

Strategies to Enhance Decision-Making in Cybersecurity

To overcome these challenges and improve decision-making in this OODA Loop Phase, organizations can adopt several strategies:

- **Develop a culture of learning:** Organizations should encourage a culture of continuous learning and improvement, where decision-makers regularly update their knowledge and skills to stay abreast of the latest cybersecurity trends and best practices.
- **Foster collaboration and diversity:** By promoting collaboration and diversity within decision-making teams, organizations can leverage a broader range of perspectives and expertise, leading to more robust and innovative solutions to cybersecurity challenges.

- **Leverage technology and analytics:** Utilizing advanced analytics tools and technologies, such as artificial intelligence and machine learning, can help organizations process and analyze vast amounts of data more efficiently, enabling them to make more informed and timely decisions. Triggers can be established to involve human intervention where needed to ensure the process is effective.
- **Implement structured decision-making processes:** Adopting structured decision-making frameworks such as the OODA Loop will provide a systematic and iterative approach to gathering, processing, and acting on information to help organizations navigate the complexities of cybersecurity decision-making more effectively.
- **Encourage critical thinking and open dialogue:** Organizations should promote critical thinking and open dialogue among decision-makers, creating an environment where dissenting opinions and alternative viewpoints are valued and considered.
- **Prioritize risk management and assessment:** By regularly conducting risk assessments and prioritizing risks based on their potential impact and likelihood, organizations can better focus their decision-making efforts on the most critical threats and vulnerabilities.
- **Develop and maintain a playbook of response strategies:** Having a playbook of predefined response strategies for various types of cyber threats can help organizations expedite decision-making in times of crisis and ensure that they are well-prepared to respond effectively to emerging threats.
- **Incorporate feedback and learning loops:** Organizations should continuously evaluate the outcomes of their decisions and integrate feedback and lessons learned into their decision-making processes to improve their ability to adapt and respond to evolving cyber threats.

Chapter Sources

1. Endsley, M. R. (1995). Toward a theory of situation awareness in dynamic systems. Human Factors, 37(1), 32-64. **https://doi.org/10.1518/001872095779049543**

2. Goodwin, G. C., & Sin, K. K. (2014). Adaptive filtering prediction and control. Courier Corporation.

3. Kahneman, D. (2011). Thinking, fast and slow. Macmillan.

4. McChrystal, S., Collins, T., Silverman, D., & Fussell, C. (2015). Team of teams: New rules of engagement for a complex world. Penguin.

5. Tversky, A., & Kahneman, D. (1974). Judgment under uncertainty: Heuristics and biases. Science, 185(4157), 1124-1131. **https://doi.org/10.1126/science.185.4157.1124**

6. Weick, K. E., & Sutcliffe, K. M. (2006). Mindfulness and the quality of organizational attention. Organization Science, 17(4), 514-524. **https://doi.org/10.1287/orsc.1060.0196**

CHAPTER 4: The Act Phase

"Action is the foundational key to all success." - Pablo Picasso

Implementation and Execution of Cybersecurity Strategies using the OODA Loop

In the Act Phase organizations put their decisions into practice, implementing and executing cybersecurity strategies that address identified threats and vulnerabilities. In the cybersecurity context, this Phase involves deploying countermeasures, adjusting security controls, and executing response plans. This Phase is critical for ensuring that the organization's cybersecurity posture is effective and continually adapts to the evolving threat landscape.

The Importance of the Act Phase

The Act Phase is essential because it operationalizes decisions, translating strategic plans into real-world actions that protect the organization's digital assets. It is worth noting that timely and effective action is crucial in the fast-paced and constantly evolving world of cybersecurity, where threats can emerge and exploit vulnerabilities in a matter of minutes or hours.

Key Components of the Act Phase

The Act Phase encompasses several key components that contribute to its success: Implementation of Countermeasures; Adjusting Security Controls; Executing Response Plans; Communications and Collaboration; and Continuous Improvement.

Implementation of Countermeasures

Organizations must deploy appropriately tailored countermeasures to mitigate the risks and vulnerabilities that are identified. Countermeasures can be technical (e.g., firewalls, intrusion detection systems, encryption), procedural (e.g., incident response plans, security policies), or human-focused (e.g., training and awareness programs). The choice of countermeasures depends on the organization's risk appetite, available resources, and the specific threats being addressed.

Adjusting Security Controls

As organizations learn more about the threat landscape and their own security posture, they must continually adjust their controls to maintain an optimal balance between security and usability. Adjustments can include tightening or loosening access controls, modifying security configurations, or reallocating resources to address new or emerging threats.

Executing Response Plans

When a security incident occurs, organizations must execute their response plans, which include designated steps for detecting, containing, eradicating, and recovering from the incident. The effectiveness of these plans often depends on the organization's ability to act quickly and decisively, minimizing the impact of the incident on its operations, reputation, and customers.

Communication and Collaboration

Effective action in the Act Phase requires clear communication and collaboration among various stakeholders, including internal teams (e.g., IT, security, legal, and public relations) and external partners (e.g., vendors, regulators, and law enforcement). Sharing information about threats, vulnerabilities, and best practices will help organizations stay abreast of the latest trends and develop more effective defense strategies.

Continuous Improvement

The Act Phase should also include a focus on continuous improvement in which organizations learn from their experiences and refine their security posture. This may involve conducting post-incident reviews to identify lessons learned, updating security policies and procedures, and investing in innovative technologies or training programs to enhance the organization's cybersecurity capabilities.

Challenges in the Act Phase

Organizations face several challenges in executing the Act Phase effectively:

Limited Resources: Many organizations struggle with limited resources, such as budget, personnel, and technical expertise, which can constrain their ability to implement and execute their cybersecurity strategies effectively. This challenge is particularly acute for small and medium-sized businesses, which often lack the dedicated cybersecurity staff and financial resources of larger organizations.

Coordination and Collaboration: The Act Phase requires close coordination and collaboration among various stakeholders, both internal and external. However, organizations often face challenges in establishing effective communication channels and fostering a culture of collaboration, particularly when dealing with sensitive security issues.

Complexity and Interdependencies: Modern organizations rely on complex, interconnected IT systems and third-party services, which can make it difficult to identify and address security vulnerabilities in a timely manner. In addition, the rapid pace of technological change and the ever-evolving threat landscape can make it challenging for organizations to keep up with the latest trends and adapt their security strategies accordingly.

Legal and Regulatory Compliance: Organizations must also navigate a complex web of legal and regulatory requirements, which can impose additional constraints on their cybersecurity strategies. Failure to comply with these requirements can result in fines, penalties, and reputational damage, making it crucial for organizations to understand and adhere to the relevant laws and regulations.

Balancing Security and Usability: Organizations often face a delicate balancing act between implementing robust security measures and ensuring that these measures do not overly burden users or impede business operations. Striking the right balance can be challenging, as overly restrictive security controls can lead to user frustration and workarounds that ultimately weaken the organization's security posture.

Best Practices for the Act Phase

To address these challenges and ensure the successful execution of the Act Phase, organizations can adopt several best practices:

Prioritize Actions Based on Risk: Given the resource constraints that many organizations face, it is crucial to prioritize actions based on their potential impact on the organization's risk profile. This process involves assessing the likelihood and potential consequences of each threat or vulnerability and allocating resources accordingly.

Foster a Culture of Collaboration: Creating a culture of collaboration can help break down silos and improve communication among stakeholders, both internal and external. This includes establishing clear channels for sharing information, promoting a sense of shared responsibility for cybersecurity, and encouraging cross-functional collaboration on security initiatives.

Adopt an Agile Approach: Given the rapidly evolving nature of the threat landscape, organizations should adopt an agile approach to cybersecurity, which emphasizes flexibility, adaptability, and continuous improvement. This includes regularly reviewing and updating security policies and procedures, investing in ongoing training and education, and staying informed about the latest trends and best practices.

Leverage Automation and Technology: Organizations can also leverage automation and technology to improve the efficiency and effectiveness of their cybersecurity efforts. This may include deploying automated tools for vulnerability scanning, intrusion detection, and incident response, as well as investing in advanced analytics and artificial intelligence (AI) solutions to enhance threat detection and decision-making capabilities.

Align with Legal and Regulatory Requirements: Finally, organizations should ensure that their cybersecurity strategies are aligned with the relevant legal and regulatory requirements. This includes staying abreast of changes in the regulatory landscape, conducting regular compliance audits, and working closely with legal counsel to ensure that security measures do not inadvertently violate laws or regulations.

By understanding the challenges and best practices associated with the Act Phase, organizations can improve their ability to respond to and mitigate cyber threats, ultimately enhancing their overall security and resilience.

Chapter Sources

1. Bayuk, J. (2010). Cyber Security Policy Guidebook. John Wiley & Sons.
2. Cichonski, P., Millar, T., Grance, T., & Scarfone, K. (2012). Computer Security Incident Handling Guide: Recommendations of the National Institute of Standards and Technology. NIST Special Publication 800-61 Revision 2. Retrieved from **https://nvlpubs.nist.gov/nistpubs/SpecialPublications/NIST.SP.800-61r2.pdf**
3. ENISA. (2016). Guidelines for SMEs on the Security of Personal Data Processing. Retrieved from **https://www.enisa.europa.eu/publications/guidelines-for-smes-on-the-security-of-personal-data-processing**
4. IBM. (2022). Cost of a Data Breach Report. Retrieved from **https://www.ibm.com/security/digital-assets/cost-data-breach-report/#/**
5. ISO/IEC. (2022). Information technology – Security techniques – Information security incident management – Part 2: Guidelines to plan and prepare for incident response. ISO/IEC 27035-2:2023. Retrieved from **https://www.iso.org/standard/62071.html**

CHAPTER 5: Decision-Making in Practice: Cybersecurity Case Studies

"Experience is not what happens to you; it's what you do with what happens to you." - Aldous Huxley

To illustrate the importance of effective decision-making in cybersecurity, in this section we examine a series of OODA application case studies relevant to cybersecurity. The first two studies provide simplified examples to help the reader acclimatize to the terminology and concepts as they apply to real-world situations.

Case Study 5.1 – Using OODA to cross a street

When crossing a busy New York street, the OODA Loop can be used to help an individual navigate through the bustling environment safely and efficiently. Here's a step-by-step explanation of how someone might utilize the OODA Loop in this scenario:

1. **Observe:** The individual would first take in all the available information about the street and its surroundings. This includes noticing the traffic signals, the flow of vehicles and pedestrians, any potential hazards, or obstacles, and assessing the general environment to identify any relevant patterns or changes.
2. **Orient:** In this Phase, the individual processes the information gathered in the observation Phase to understand the situation. They would assess their own position relative to the street, recognize any risks or opportunities, and consider their previous experiences with similar environments. Additionally, they may consider any cultural norms, such as waiting for the pedestrian signal before crossing or looking both ways before stepping onto the street.

3. **Decide:** Based on the information and understanding from the previous Phases, the individual would decide about when and how to cross the busy street. Factors to consider include the safest route, the appropriate timing given the traffic flow, and whether it's better to wait for a more favorable condition or cross immediately. The decision will depend on the individual's assessment of risks and their confidence in successfully navigating the situation.

4. **Act:** Finally, the individual would execute their chosen course of action, such as waiting for the pedestrian signal and crossing at the designated crosswalk. They may also need to adjust on-the-fly while crossing, such as speeding up or slowing down to avoid oncoming traffic or changing their path to bypass any unexpected obstacles.

Throughout the entire process, the individual would continuously cycle through the OODA Loop, constantly observing, orienting, deciding, and acting as they navigate the busy New York street. This approach allows them to remain adaptive and responsive to the dynamic environment, improving their chances of crossing the street safely and effectively.

Case Study 5.2 – Choosing a restaurant

Let's consider a scenario where someone is trying to decide which restaurant to eat at with a group of friends.

1. **Observe:** The individual begins by gathering information about available restaurant options in the area. They might use online resources like restaurant review websites, social media, or mobile apps to find information about different restaurants, their menus, prices, and customer ratings. They also consider the preferences, dietary restrictions, and opinions of their friends.

2. **Orient:** In this Phase, the individual processes the information collected during the observation Phase. They consider their previous experiences with similar restaurants,

their knowledge of their friends' preferences, and the overall context of the situation (e.g., time constraints, budget, or occasion). They might also assess the pros and cons of each option, weighing factors like distance, ambiance, and the variety of dishes available.

3. **Decide:** After evaluating the information gathered and considering the relevant factors, the individual decides on which restaurant to choose. This decision may involve prioritizing certain criteria, such as selecting a restaurant that can accommodate specific dietary needs, has the best customer reviews, or fits within the group's budget.

4. **Act:** The individual communicates their decision to their friends and proceeds to make a reservation or head to the chosen restaurant. While en route, they remain alert for any added information that might impact their decision, such as discovering that the restaurant is unexpectedly closed or overcrowded.

As the individual and their friends experience the chosen restaurant, they continue to cycle through the OODA Loop, observing the service, food quality, and atmosphere, orienting this new information with their expectations, and making any necessary adjustments (e.g., ordering additional dishes or changing seats). This continuous application of the OODA Loop helps the individual adapt to the situation, make informed decisions, and ensure a satisfying dining experience for the group.

Case Study 5.3 – Proactive Cybersecurity Decision-Making using the OODA Loop

In this case study, we examine a medium-sized financial services organization that has adopted the OODA Loop as part of its cybersecurity strategy. The organization processes sensitive financial information daily and is a prime target for cybercriminals. Aware of the risks, the organization has prioritized a proactive cybersecurity approach to protect its digital assets and maintain its reputation.

1. **Observe:** The organization has a dedicated cybersecurity team that continuously monitors the network for potential threats and vulnerabilities. By employing state-of-the-art intrusion detection systems, threat intelligence feeds, and SIEM (Security Information and Event Management) tools, they can quickly detect anomalies and potential issues. During routine monitoring, the cybersecurity team identifies a potential vulnerability in one of the organization's critical applications. This vulnerability could be exploited by threat actors to gain unauthorized access to sensitive customer data.

2. **Orient:** Upon discovering the vulnerability, the cybersecurity team quickly assesses the potential risk associated with the issue. They gather information on the vulnerability from various sources, such as vulnerability databases, industry reports, and cybersecurity forums. They also analyze historical data on similar incidents to understand the potential impact and likelihood of exploitation. The cybersecurity team considers the organization's unique context, including its risk tolerance, regulatory requirements, and potential reputational damage. After thorough analysis, they determine that the vulnerability poses a high risk and requires immediate attention.

3. **Decide:** Based on the insights gained during the Orient Phase, the organization's cybersecurity team evaluates several alternatives to address the identified vulnerability. These options may include applying a security patch provided by the application vendor, implementing additional security controls, or temporarily disabling the affected application feature. After weighing the pros and cons of each alternative, the cybersecurity team decides to apply the security patch, as it is the most effective and least disruptive solution. This decision is aligned with the

organization's objectives of minimizing potential risks while maintaining business continuity.

4. **Act:** The cybersecurity team quickly communicates their decision to relevant stakeholders, such as IT operations and application development teams. They collaborate to plan and execute the patch deployment, ensuring minimal disruption to the organization's operations. Once the patch is deployed, the cybersecurity team verifies that the vulnerability has been successfully mitigated. They also update their incident response procedures and vulnerability management processes based on the lessons learned from this case.

This case study demonstrates the effectiveness of using the OODA Loop in cybersecurity decision-making. By proactively monitoring and assessing risks, the organization was able to identify a critical vulnerability, evaluate potential courses of action, and make a strategic decision to mitigate the risk before it could be exploited by cybercriminals. This proactive approach not only protected the organization's digital assets but also helped maintain its reputation in the highly competitive financial services industry.

Case Study 5.4 – The Consequences of Ineffective Decision-Making in Cybersecurity

In this case study, we explore a small e-commerce company that has not yet implemented a structured decision-making framework like the OODA Loop for its cybersecurity program. The company deals with online sales and processes customer transactions daily, making it an attractive target for cybercriminals.

1. **Observe:** The e-commerce company has a small IT team that handles various responsibilities, including cybersecurity. Due to limited resources and a lack of specialized cybersecurity expertise, the team struggles to monitor and detect potential threats and vulnerabilities effectively. As a result, a critical vulnerability in the company's web application goes unnoticed.

2. **Orient:** An external threat actor discovers the vulnerability and successfully exploits it to gain unauthorized access to the company's customer database. The company becomes aware of the breach only when customers start reporting unauthorized transactions on their accounts. Once the breach is detected, the IT team scrambles to assess the situation and determine the potential impact on the company. However, they lack the experience and tools to conduct a comprehensive analysis quickly, leading to a delay in understanding the full scope of the breach and its implications.

3. **Decide:** The e-commerce company's management team struggles to decide on the best course of action to address the breach. They are overwhelmed by the complexity and urgency of the situation and lack a structured decision-making framework to guide their response. Due to a lack of accurate information, effective communication, and a clear understanding of the potential consequences, the company's management makes several suboptimal decisions. These decisions include attempting to downplay the breach's severity, delaying the notification of affected customers, and failing to prioritize remediation efforts. Their approach resulted in ineffective decision-making.

4. **Act:** As a result of the company's slow and ineffective response, the breach continues to impact more customers. Word of the breach spreads quickly, and the company's reputation suffers. Regulators become involved, and the company faces potential fines and legal action for failing to protect customer data adequately. Eventually, the company is forced to invest significant resources in incident response, remediation efforts, and public relations damage control. They also need to implement new security measures and processes to prevent future breaches, all of which could have been avoided with more proactive and effective decision-making.

This case study highlights the consequences of ineffective decision-making in cybersecurity. The e-commerce company's lack of a structured decision-making framework like the OODA Loop led to a delayed response and poor choices when dealing with a critical breach. The company suffered financial losses, reputational damage, and potential legal and regulatory consequences.

In contrast, adopting a proactive cybersecurity approach guided by the OODA Loop could have enabled the company to identify the vulnerability, assess the risk, and make timely and informed decisions to mitigate the threat before it could be exploited. This study illustrates how implementing structured decision-making processes like the OODA Loop is essential to navigating the complex and rapidly evolving cybersecurity landscape effectively.

Case Study 5.5 – Rapid Response to a Ransomware Attack using the OODA Loop

In this case study, we examine a large healthcare organization that has implemented the OODA Loop as part of its cybersecurity incident response strategy. Healthcare organizations often hold vast amounts of sensitive patient data, making them prime targets for cybercriminals. Ransomware attacks on such organizations can have devastating consequences on patient care and the organization's operations.

1. **Observe:** The healthcare organization has a robust cybersecurity program in place, including continuous monitoring and threat intelligence capabilities. In the normal course of business, the organization's cybersecurity team receives an alert from their endpoint detection and response (EDR) solution, indicating a potential ransomware attack. The EDR solution has detected unusual file encryption activity on one of the organization's file servers, suggesting that a ransomware attack is in progress. The cybersecurity team immediately begins an investigation to confirm the attack and assess its scope.

2. **Orient:** Upon confirming the ransomware attack, the cybersecurity team quickly gathers information about the

specific strain of ransomware involved, the extent of the affected systems, and the potential impact on the organization's operations. The team reviews relevant threat intelligence reports, ransomware mitigation best practices, and historical data on similar incidents to understand the potential consequences and identify effective countermeasures. They also consider the organization's unique context, such as its risk tolerance, regulatory requirements, and the potential impact on patient care.

3. **Decide:** Armed with the insights gathered during the Orient Phase, the organization's cybersecurity team evaluates several options to address the ransomware attack. These options include tactics that may be used in tandem, such as isolating the affected systems to prevent further spread, attempting to restore from backups, engaging external experts for assistance, and negotiating with the threat actors. Considering the potential harm to patient care and the organization's operations, the cybersecurity team decides to prioritize containment and recovery efforts. They opt to isolate the affected systems, restore them from backups, and engage external experts to support the investigation and recovery process.

4. **Act:** The cybersecurity team quickly communicates their decision to relevant stakeholders, such as IT operations, incident response teams, and senior management. They work together to implement the containment measures, isolate the affected systems, and initiate the recovery process using available backups.

Concurrently, the external experts join the investigation to help identify the ransomware's infection vector and develop additional mitigation strategies to prevent future incidents. The organization also communicates the incident to relevant authorities, patients, and partners to maintain transparency and comply with regulatory requirements.

This case study demonstrates the value of using the OODA Loop in being effective in responding to a ransomware attack. By quickly observing, orienting, deciding, and acting, the healthcare organization managed to contain the ransomware attack, minimize the impact on patient care, and expedite the recovery process.

The organization's proactive adoption of the OODA Loop in its cybersecurity incident response strategy allowed them to make informed, strategic decisions under pressure, highlighting the importance of structured decision-making processes in managing cyber threats.

Case Study 5.6 – Insider Threat Detection and Mitigation using the OODA Loop

In this case study, we explore a multinational corporation that has implemented the OODA Loop as part of its cybersecurity strategy to address insider threats. Insider threats are a significant concern for organizations of all sizes, as they can involve employees, contractors, or business partners who have authorized access to sensitive information and can cause substantial damage, either intentionally or inadvertently.

1. **Observe:** The multinational corporation has implemented a comprehensive cybersecurity program that includes monitoring user activities to detect and prevent insider threats. They have deployed user behavior analytics (UBA) tools to identify anomalies and potential risks associated with insider activities. During routine monitoring, the organization's cybersecurity team detects an employee accessing and downloading an unusually large volume of sensitive company data, raising suspicion of potential data exfiltration.

2. **Orient:** The cybersecurity team promptly begins an investigation to determine the potential risk associated with the detected anomaly. They gather information about the employee's role, job responsibilities, and access privileges to understand whether the observed activities align with their normal duties. Additionally, the team reviews historical data and compares the employee's activities with their peers to identify any inconsistencies or red flags. After thorough

analysis, they determine that the employee's actions are suspicious and potentially harmful to the organization.

3. **Decide:** Based on the insights gained during the Orient Phase, the organization's cybersecurity team evaluates several options to address the potential insider threat. These options include monitoring the employees' activities more closely, revoking or restricting their access to sensitive information, conducting a more detailed investigation, or involving the human resources department. The cybersecurity team decides to conduct a more detailed investigation while simultaneously implementing additional monitoring and access controls to limit the employee's access to sensitive data. This approach aims to minimize the risk of data exfiltration while gathering more information to make an informed decision.

4. **Act:** The cybersecurity team quickly communicates their decision to relevant stakeholders, including IT operations, human resources, and senior management. They collaborate to implement the additional monitoring and access controls and initiate the in-depth investigation. Upon completing the investigation, the team finds evidence that the employee was indeed attempting to exfiltrate sensitive company data for personal gain. The organization takes appropriate disciplinary action, including termination of employment, revocation of privileged access, and reporting the incident to law enforcement.

This case study highlights the importance of using a structured decision-making framework like the OODA Loop in addressing insider threats. By proactively observing, orienting, deciding, and acting, the multinational corporation managed to identify a potential insider threat, assess the risk, and take appropriate actions to mitigate the risk before any grave damage occurred.

The organization's adoption of the OODA Loop in its cybersecurity strategy demonstrates the value of a proactive and structured approach in managing complex and often elusive threats, such as insider threats.

Case Study 5.7 – Enhancing Security Posture through the OODA Loop in a Smart City

In this case study, we explore how a smart city uses the OODA Loop to enhance its cybersecurity posture. Smart cities leverage modern technologies, such as the Internet of Things (IoT) and cloud computing, to provide better services and improve the quality of life for their citizens. However, these technological advancements also introduce new cybersecurity risks and vulnerabilities that must be addressed.

1. **Observe:** The smart city has implemented a robust cybersecurity program, which includes monitoring and managing the security of its various interconnected systems. The city's cybersecurity team continuously collects data from sensors, IoT devices, and other critical infrastructure components to detect potential threats and vulnerabilities. During a routine network scan, the team identifies a vulnerability in the city's traffic management system that could potentially be exploited by cybercriminals to disrupt traffic flow or cause accidents.

2. **Orient:** The cybersecurity team immediately begins gathering information about the vulnerability, including its potential impact, the likelihood of exploitation, and the availability of patches or other mitigation techniques. The team also considers the city's unique context, such as the potential consequences of a successful attack on the traffic management system and the city's risk tolerance. In addition to analyzing the vulnerability, the team examines historical data on similar incidents and relevant threat intelligence reports to better understand the threat landscape and identify effective countermeasures.

3. **Decide:** After analyzing the information gathered during the Orient Phase, the city's cybersecurity team evaluates several options to address the vulnerability. These

options include patching the vulnerability, implementing additional security measures, or temporarily disabling affected services. Considering the potential impact on public safety and the city's operations, the team decides to prioritize patching the vulnerability and enhancing the security of the traffic management system to prevent potential exploitation.

4. **Act:** The cybersecurity team quickly communicates their decision to relevant stakeholders, such as the traffic management department, IT operations, and city management. They collaborate to implement the necessary patches and additional security measures to protect the traffic management system from potential cyberattacks. The team also reviews and updates the city's incident response plan to ensure that it adequately addresses potential threats to critical infrastructure components like the traffic management system.

This case study illustrates the value of using the OODA Loop in enhancing the cybersecurity posture of a smart city. By proactively observing, orienting, deciding, and acting, the city managed to identify a critical vulnerability, assess the associated risks, and take timely and effective actions to mitigate potential threats.

The city's adoption of the OODA Loop in its cybersecurity strategy highlights the importance of a structured decision-making process in managing the complex and ever-evolving cybersecurity landscape in smart cities.

Case Study 5.8 – Addressing GPT-based Threats in a Financial Institution using the OODA Loop

In this case study, we explore how a financial institution uses the OODA Loop to address potential cybersecurity threats posed by GPT-based technologies. Generative Pre-trained Transformers (GPT) are a type of artificial intelligence that can generate highly convincing and coherent text based on given prompts. While GPT has many legitimate applications, it also introduces new cybersecurity risks, such as social engineering, deep fakes, and other forms of misinformation.

1. **Observe:** The financial institution has a comprehensive cybersecurity program that includes monitoring and threat intelligence capabilities. The institution's cybersecurity team becomes aware of the growing trend of GPT-based threats and begins to monitor for any signs of such threats targeting their organization. During their monitoring efforts, the team detects a series of highly targeted phishing emails being sent to the institution's employees. These phishing emails appear to be generated using GPT-based technologies, making them highly convincing and difficult to detect by traditional security measures.

2. **Orient:** Upon detecting the GPT-based phishing campaign, the cybersecurity team quickly gathers information about the specific tactics, techniques, and procedures (TTPs) used by the attackers. They analyze the content of the phishing emails, the targeted employees, and the potential impact on the organization's operations. The team reviews relevant threat intelligence reports, industry best practices, and historical data on similar incidents to understand the potential consequences and identify effective countermeasures. They also consider the organization's unique context, such as its risk tolerance, regulatory requirements, and the potential impact on customer trust and reputation.

3. **Decide:** Based on the insights gathered during the Orient Phase, the financial institution's cybersecurity team evaluates several options to address the GPT-based phishing campaign. These options include improving employee awareness and training, implementing advanced email filtering and detection technologies, and engaging external experts for

assistance. Considering the potential harm to the organization's reputation and customer trust, the cybersecurity team decides to prioritize employee awareness and training, implement advanced email filtering technologies, and engage external experts to support the investigation and improve the organization's defenses against GPT-based threats.

4. **Act:** The cybersecurity team quickly communicates their decision to relevant stakeholders, such as human resources, IT operations, and senior management. They work together to enhance the organization's security awareness training program, focusing on GPT-based threats and social engineering tactics. They also implement advanced email filtering technologies that are more effective at detecting GPT-generated phishing emails. Concurrently, the external experts join the investigation to help identify the source of the GPT-based phishing campaign and develop additional mitigation strategies to prevent future incidents. The organization also communicates the incident to relevant authorities and partners to raise awareness and collaborate on addressing GPT-based threats in the financial industry.

This case study demonstrates the value of using the OODA Loop in responding to emerging and sophisticated cybersecurity threats, such as those posed by GPT-based technologies. By quickly observing, orienting, deciding, and acting, the financial institution managed to identify and mitigate the GPT-based phishing campaign, enhancing its cybersecurity posture and protecting customer trust.

The organization's proactive adoption of the OODA Loop in its cybersecurity strategy allowed them to make informed, strategic decisions under pressure, highlighting the importance of structured decision-making processes in managing novel cyber threats.

Case Study 5.9 – Utilizing the OODA Loop to Choose an AI Model for an E-commerce Business

This study examines an e-commerce company, eShops Tomorrow, which is looking to improve its customer experience and streamline its operations. The company's management team decides to explore the potential of implementing an AI model to achieve these goals. They opt to use the OODA Loop as a structured decision-making process to guide their selection and implementation of the most suitable AI model.

1. **Observe:** The management team begins by gathering data on various AI models available in the market, as well as researching how similar companies have successfully employed AI to enhance their businesses. They collect information on AI-powered chatbots, recommendation engines, and inventory management systems. The team also identifies the company's key performance indicators (KPIs), such as conversion rates, customer satisfaction, and revenue, which will be used to gauge the value of models and measure the success of the AI implementation.

2. **Orient:** With the collected data, the team analyzes the potential benefits and drawbacks of each AI model, considering factors such as cost, ease of integration, and scalability. They involve the IT department and key stakeholders to ensure that the chosen AI model aligns with the company's technological capabilities and overall business strategy. The team also reviews the company's current processes and identifies areas where AI could have the most significant impact, such as customer support, product recommendations, and inventory management.

3. **Decide:** After thoroughly analyzing the data, the management team decides that a combination of an AI-powered chatbot and a recommendation engine would provide the most value to the company. The chatbot would improve customer support by providing instant responses to common queries, while the recommendation engine would

enhance the shopping experience by suggesting personalized product recommendations based on customer preferences and browsing behavior. The team chooses specific AI models that are cost-effective, scalable, and compatible with the company's existing technological infrastructure.

4. **Act:** The management team proceeds with the implementation of the chosen AI models. They collaborate with the IT department and the AI solution providers to integrate the chatbot and recommendation engine into the company's website and mobile app. The team also conducts training sessions for customer support staff to familiarize them with the new chatbot system and to ensure that they can handle more complex customer inquiries effectively. Throughout the implementation process, the team closely monitors the performance of the AI models, tracking their predetermined KPIs. They use the feedback obtained from customers and staff to make necessary adjustments to the AI models, ensuring that they are continuously optimized for the best possible results.

By using the OODA Loop, eShop Tomorrow's management team was able to make an informed decision on choosing the most suitable AI models for their business. They successfully implemented an AI-powered chatbot and recommendation engine that improved customer support and personalized the shopping experience, leading to higher customer satisfaction and increased revenue. This case study demonstrates the value of using the OODA Loop as a structured decision-making process for selecting and implementing AI models in a business setting.

Chapter Sources

Case Study 5.3 - Detecting and Mitigating Advanced Persistent Threats (APTs) in a Large Corporation

FireEye. (2021). M-Trends 2022 Metrics, Insights and Guidance from the Front Lines. Retrieved from **https://www.mandiant.com/resources/blog/m-trends-2022**

SANS Institute. (2020). Cyber Threat Intelligence: A Practical Guide to Implementing an Intelligence-Driven Security Program. Retrieved from **https://www.sans.org/reading-room/whitepapers/threats/cyber-threat-intelligence-36357**

Case Study 5.4 - Implementing the OODA Loop in a Small Business for Proactive Cyber Defense

NIST. (2016). Small Business Information Security: The Fundamentals. Retrieved from **https://nvlpubs.nist.gov/nistpubs/ir/2016/NIST.IR.7621r1.pdf**

DHS. (2021). Cyber Security Evaluation Tool (CSET) for Small Businesses. Retrieved from **https://www.cisa.gov/cyber-resource-hub**

Case Study 5.5 - Ransomware Attack on a Healthcare Organization

Verizon. (2021). Data Breach Investigations Report. Retrieved from **https://www.verizon.com/business/resources/reports/dbir/**

NIST. (2018). Framework for Improving Critical Infrastructure Cybersecurity. Retrieved from **https://nvlpubs.nist.gov/nistpubs/CSWP/NIST.CSWP.04162018.pdf**

Case Study 5.6 - Insider Threat Detection and Mitigation using the OODA Loop

Gartner. (2021). Market Guide for User and Entity Behavior Analytics. Retrieved from **https://www.gartner.com/en/documents/3996019/market-guide-for-user-and-entity-behavior-analytics**

CERT. (2018). Common Sense Guide to Mitigating Insider Threats. Retrieved from

https://resources.sei.cmu.edu/library/asset-view.cfm?assetid=539423
Case Study 5.7 - Enhancing Security Posture through the OODA Loop in a Smart City
ENISA. (2020). Smart Cities Security Recommendations. Retrieved from **https://www.enisa.europa.eu/publications/smart-cities-security-recommendations**
ISO. (2021). Smart Cities - Information Security Management Systems. Retrieved from **https://www.iso.org/standard/73439.html**
Case Study 5.8 - Addressing GPT-based Threats in a Financial Institution using the OODA Loop
FSB. (2020). Effective Practices. For Cyber Incident Response and Recovery Retrieved from **https://www.fsb.org/2020/10/effective-practices-for-cyber-incident-response-and-recovery-final-report/**

CHAPTER 6: Feed Forward and Feed Backwards

"The only way to learn from your mistakes is to admit them rather than blame them on someone else." - John C. Maxwell

The OODA Loop is a powerful decision-making model that has gained widespread recognition and application across various domains, including cybersecurity. An essential part of understanding the OODA Loop and its value in cybersecurity lies in examining the concepts of feed forward and feedback mechanisms within the loop. By delving into these concepts, we can better appreciate their significance and how they contribute to the success of cybersecurity programs.

Feed forward and feedback are two complementary mechanisms in the OODA Loop that facilitate the effective flow of information and learning throughout the decision-making process. Feed forward refers to the process of using past experiences, predictions, and proactive measures to shape and influence future actions and outcomes, whereas feedback involves the evaluation and analysis of the results of past actions to inform and refine future decisions. These mechanisms play a crucial role in enabling organizations and cybersecurity professionals to continuously learn, adapt, and improve their strategies and tactics in the face of ever-evolving cyber threats.

To fully appreciate the concepts of feed forward and feedback in the context of the OODA Loop and cybersecurity, a fundamental understanding of the underlying principles of the OODA Loop is required.

- The OODA Loop consists of four interrelated Phases: Observe, Orient, Decide, and Act.
- These Phases form a continuous loop that enables individuals and organizations to gather information, analyze and interpret it, make decisions, and take appropriate actions in response to dynamic and uncertain situations.

- The OODA Loop emphasizes the importance of speed, adaptability, and agility in decision-making, particularly in complex and rapidly changing environments such as the cyber domain.

The value of the feed forward and feedback mechanisms in the OODA Loop cannot be overstated in the realm of cybersecurity. As cyber threats continue to grow in complexity and sophistication, organizations must adopt a proactive and adaptive approach to safeguard their digital assets, networks, and systems. By leveraging the feed forward and feedback mechanisms, cybersecurity professionals can better anticipate, respond to, and learn from the ever-evolving cyber threat landscape, ultimately enhancing their organization's security posture and resilience.

The feed forward mechanism in the OODA Loop is particularly evident in the Orient Phase, which encompasses the processes of analysis and synthesis, cultural traditions, genetic heritage, latest information, and previous experiences. Through the Orient Phase, individuals and organizations can proactively shape their perceptions, assumptions, and mental models based on past experiences, knowledge, and expectations, thereby influencing their subsequent decisions and actions. In cybersecurity, the feed forward mechanism is critical in anticipating potential threats, vulnerabilities, and risks, as well as in developing proactive strategies and measures to protect an organization's digital assets, networks, and systems.

For instance, a cybersecurity professional may leverage their previous experiences with certain threat actors, attack vectors, or security incidents to predict and anticipate similar threats in the future. By doing so, they can proactively develop and implement security measures, such as updating firewalls, patching software vulnerabilities, or enhancing user awareness and training programs, to mitigate the risk of future attacks.

Additionally, the feed forward mechanism can be instrumental in fostering a culture of continuous learning and improvement within an organization, as it encourages individuals and teams to reflect on their past experiences, learn from them, and apply these lessons to enhance their future performance.

On the other hand, the feedback mechanism in the OODA Loop is closely related to the Act Phase and the subsequent return to the Observe Phase. Feedback involves the evaluation and analysis of the outcomes and consequences of past actions, allowing individuals and organizations to learn from their successes and failures and refine their decision-making processes accordingly. In the context of cybersecurity, the feedback mechanism is essential in ensuring that security strategies, policies, and measures remain effective and relevant in the face of an ever-changing threat landscape.

For example, following a cyber incident, a cybersecurity team may conduct a post-incident review to evaluate the effectiveness of their response actions, identify lessons learned, and develop recommendations for improvement. The feedback mechanism enables the team to continuously learn from their experiences, identify areas for improvement, and refine their response strategies and tactics over time. Moreover, the feedback mechanism can help organizations identify trends, patterns, and systemic issues in their cybersecurity programs, allowing them to address these issues proactively and enhance their overall security posture.

One of the key benefits of the feed forward and feedback mechanisms in cybersecurity is their ability to support continuous learning and improvement. By proactively analyzing past experiences, anticipating future threats, and evaluating the results of past actions, organizations can continually refine their security strategies, policies, and measures to stay ahead of the curve. This iterative process of learning, adapting, and improving is essential in the face of an increasingly dynamic and uncertain cyber environment.

Another significant advantage of the feed forward and feedback mechanisms is their capacity to foster a culture of collaboration, communication, and shared responsibility within an organization. As cybersecurity professionals engage in the processes of analysis, synthesis, evaluation, and learning, they can break down silos, share knowledge and insights, and work together more effectively to address emerging threats and vulnerabilities. This collaborative approach not only strengthens the organization's overall security posture; it also empowers individuals and teams to take ownership of their roles in protecting the organization's digital assets, networks, and systems.

Moreover, the feed forward and feedback mechanisms enable organizations to better align their cybersecurity programs with their business objectives, risk appetite, and regulatory requirements. By proactively identifying and addressing potential threats, vulnerabilities, and risks, organizations can optimize their resource allocation, prioritize their security investments, and demonstrate their commitment to cybersecurity to stakeholders, regulators, and customers. Additionally, the feedback mechanism can highlight relevant observations and metrics to support ongoing risk management, compliance, and governance efforts, further enhancing the organization's reputation, trustworthiness, and competitiveness in the marketplace.

The feed forward and feedback mechanisms within the OODA Loop offer immense value to cybersecurity professionals and organizations alike. By embracing these mechanisms, organizations can develop a more proactive, adaptive, and resilient approach to cybersecurity, continuously learning from their past experiences and leveraging this knowledge to shape and inform their future decisions and actions. In an era of rapidly evolving cyber threats and increasing reliance on digital technologies, the feed forward and feedback mechanisms within the OODA Loop are more critical than ever to ensuring the security, stability, and success of organizations worldwide.

Case Study 6.1 – Improving the OODA Loop Through Feedback in a Financial Services Company

A large financial services company had invested significantly in its cybersecurity infrastructure and had implemented an incident response plan to address potential cyber threats. The company faced a wide range of cyber risks, including phishing attacks, ransomware, and advanced persistent threats targeting its sensitive financial data.

The company experienced a sophisticated spear-phishing attack targeting its finance department employees. The attackers sent well-crafted emails, appearing to be from trusted internal sources, with malicious attachments. Upon opening the attachments, a few employees unknowingly initiated the installation of malware on their systems.

The security operations center (SOC) team detected unusual activities on the network and promptly initiated the incident response plan. They identified and isolated the affected systems, removed the malware, and conducted a thorough investigation to understand the extent of the breach. Fortunately, the damage was limited, and no sensitive data was exfiltrated.

Feedback Process

Following the incident, the company conducted a comprehensive post-incident review to analyze the effectiveness of their response and identify areas for improvement. The review revealed several gaps in the incident response process, particularly in terms of communication, employee awareness, and the speed of response.

1. **Communication:** The review identified that communication between the SOC team and the finance department was slow and inefficient, delaying the response to the phishing attack.

2. **Employee Awareness:** The investigation found that the employees in the finance department had not received sufficient training on how to identify and report phishing emails.

3. **Speed of Response:** The review revealed that the SOC team's initial response to the incident was slower than anticipated, highlighting the need for faster decision-making and action.

Improvements

Based on the feedback from the post-incident review, the company implemented several improvements to its incident response plan and overall cybersecurity program:

1. **Improved communication:** The company established a streamlined communication process between the SOC team and other departments, ensuring faster information sharing during incidents. They also implemented a centralized incident reporting system to facilitate quicker identification and response to potential threats.

2. **Enhanced employee awareness:** The company conducted targeted security awareness training for the finance department employees, focusing on phishing attacks and other relevant threats. They also implemented a company-wide security awareness program to improve employees' understanding of cyber risks and their role in preventing incidents.

3. **Faster decision-making and action:** The company revised its incident response plan to include clear guidelines and protocols for faster decision-making during incidents. They also conducted regular incident response drills to ensure that the SOC team was well-prepared to respond quickly and effectively to potential threats.

Outcome

The improvements made based on the feedback from the spear-phishing incident significantly strengthened the company's overall cybersecurity posture. A few months later, the company faced another targeted phishing attack. This time, however, the finance department employees quickly identified and reported the suspicious emails to the SOC team. Thanks to the improved communication process and faster decision-making, the SOC team was able to respond promptly and prevent the installation of malware on the company's systems.

By leveraging the feedback mechanism in the OODA Loop, the financial services company successfully enhanced its cybersecurity program, preventing a potentially damaging incident. The case study highlights the value of continuous learning and improvement through feedback, ultimately leading to a more resilient and effective cybersecurity posture.

CHAPTER 7: Implicit Guidance and Control

"Intuition will tell the thinking mind where to look next." - Jonas Salk

Implicit Guidance and Control in the OODA Loop within Cybersecurity

The OODA Loop has been a widely acknowledged decision-making framework in various domains, including military strategy, business management, and more recently, cybersecurity. Quickly reviewing, the OODA Loop consists of four Phases: Observe, Orient, Decide, and Act. While each Phase is essential to the overall effectiveness of the process, the concept of Implicit Guidance and Control plays a significant role in the OODA Loop's application within the realm of cybersecurity. This concept refers to the innate ability of an individual or organization to make rapid decisions and take appropriate actions without relying solely on explicit instructions or predefined procedures. Implicit Guidance and Control are intricately linked to the orientation Phase of the OODA Loop and are influenced by factors such as cultural traditions, genetic heritage, previous experiences, added information, and analysis and synthesis.

In the context of cybersecurity, the importance of Implicit Guidance and Control cannot be overstated. It is crucial for organizations and cybersecurity professionals to cultivate the ability to make quick, informed decisions and adapt their security strategies and tactics based on the unique characteristics of each situation. Cyber threats are constantly evolving, and organizations face an increasingly complex and dynamic threat landscape. As such, traditional, rule-based security measures and explicit guidance are often insufficient to keep pace with the rapidly changing nature of cyber threats.

One of the key factors influencing Implicit Guidance and Control in the OODA Loop is cultural traditions. In the cybersecurity context, cultural traditions can refer to the established norms, values, and practices within an organization, as well as the broader security community. These traditions shape the way individuals perceive and interpret information, ultimately influencing their decision-making processes. For example, an organization that emphasizes a proactive approach to security will likely prioritize threat hunting, incident response, and continuous monitoring. By understanding and leveraging cultural traditions, cybersecurity professionals can better align their actions with the organization's overall security strategy and foster a culture that supports effective decision-making and rapid response to threats.

Another factor influencing Implicit Guidance and Control in the OODA Loop is genetic heritage. While genetic heritage may not have a direct impact on cybersecurity, it can still influence individuals' cognitive processes, risk tolerance, and decision-making styles. By acknowledging these inherent differences, cybersecurity professionals can better understand their own decision-making tendencies and adapt their approach to the unique demands of the cyber environment.

Previous experience also plays a critical role in shaping Implicit Guidance and Control within the OODA Loop. By learning from previous experiences and integrating these lessons into their decision-making processes, cybersecurity professionals can enhance their ability to anticipate and respond to emerging threats effectively. In cybersecurity, individuals and organizations continuously accumulate experiences through their encounters with various threats, incidents, and security challenges. These experiences inform their understanding of the threat landscape, shape their perception of risk, and guide their decision-making processes. For example, an organization that has experienced a ransomware attack may prioritize backup and recovery capabilities, while one that has faced persistent advanced persistent threat (APT) campaigns may focus on threat intelligence and network segmentation.

New information is another crucial factor that influences Implicit Guidance and Control in the OODA Loop. In the dynamic world of cybersecurity, staying informed about the latest threats, vulnerabilities, and technological developments is essential for effective decision-making. Cybersecurity professionals must continuously update their knowledge and skills, incorporating the latest information into their decision-making processes and adapting their strategies and tactics accordingly. By maintaining a high level of situational awareness and staying abreast of the latest trends and developments in cybersecurity, professionals can make more informed decisions and better anticipate and respond to emerging threats.

Analysis and synthesis are also essential components of Implicit Guidance and Control within the OODA Loop. In the cybersecurity context, analysis refers to the process of examining and interpreting data, information, and intelligence to gain insights into the threat landscape, vulnerabilities, and potential risks. Synthesis, on the other hand, involves combining these insights with existing knowledge and experiences to develop a comprehensive understanding of the situation and formulate appropriate strategies and actions. Both analysis and synthesis are critical to the effective application of the OODA Loop in cybersecurity, as they enable professionals to make sense of the vast amounts of data and information they encounter daily and convert them into actionable intelligence.

For example, a cybersecurity analyst may analyze logs from a network intrusion detection system (IDS) to identify patterns of malicious activity, while also synthesizing this information with threat intelligence feeds and historical incident data to determine the most likely threat actors and their motives. This process of analysis and synthesis enables the analyst to make informed decisions about the organization's security posture, allocate resources effectively, and develop targeted responses to specific threats.

To fully leverage the power of Implicit Guidance and Control in the OODA Loop, organizations, and cybersecurity professionals should focus on the following key areas:

- **Cultivate a security culture:** Foster a security-focused culture that values adaptability, continuous learning, and proactive threat management. This will help to create an environment in which Implicit Guidance and Control can thrive.

- **Develop situational awareness:** Continuously monitor the threat landscape, stay informed about the latest trends and developments in cybersecurity, and maintain an elevated level of situational awareness to support effective decision-making.
- **Learn from experience:** Emphasize the importance of learning from previous experiences, both positive and negative, and integrating these lessons into the decision-making process.
- **Enhance analytical and synthesis capabilities:** Invest in the development of analytical and synthesis skills among cybersecurity professionals and provide them with the tools and resources they need to effectively analyze and interpret data, information, and intelligence.
- **Encourage diversity:** Recognize the value of diversity in decision-making and promote the inclusion of different perspectives, experiences, and cognitive styles within the cybersecurity team.
- **Implement feedback loops:** Establish feedback mechanisms to facilitate continuous improvement and learning, enabling cybersecurity professionals to refine their decision-making processes and adapt their strategies and tactics over time.
- **Focus on adaptability and agility:** Prioritize adaptability and agility in cybersecurity strategies, ensuring that the organization can rapidly respond to and recover from emerging threats and incidents.

By embracing these principles and focusing on the factors that influence Implicit Guidance and Control within the OODA Loop, organizations and cybersecurity professionals can enhance their decision-making capabilities, improve their ability to anticipate and respond to emerging threats, and ultimately, strengthen their overall security posture.

Case Study 7.1 – Speeding Up the OODA Loop with Implicit Guidance and Control

A global financial services company faced a sophisticated cyber-attack targeting its internal network and customer data. This case study highlights how the organization's investment in fostering Implicit Guidance and Control significantly sped up its OODA Loop, leading to a faster resolution of the incident and a clear return on investment.

The financial services company had a robust cybersecurity program in place, encompassing advanced threat intelligence, continuous monitoring, and proactive incident response capabilities. Recognizing the importance of adaptability and agility in responding to emerging cyber threats, the organization had previously focused on cultivating a culture of Implicit Guidance and Control within its cybersecurity team. This approach emphasized the value of rapid decision-making, adaptability, and the integration of diverse perspectives, experiences, and cognitive styles.

Incident

The organization detected an ongoing cyber-attack, which appeared to be a sophisticated advanced persistent threat (APT) campaign. The attackers had already compromised several internal systems and were attempting to exfiltrate sensitive customer data. Time was of the essence, as the longer the attackers remained undetected, the greater the potential damage.

Response

Leveraging their Implicit Guidance and Control capabilities, the cybersecurity team rapidly assessed the situation and initiated their response. Key steps taken included:

1. **Immediate analysis and synthesis:** The team quickly analyzed the available data, including IDS logs, threat intelligence feeds, and previous incident records, synthesizing this information to develop a comprehensive understanding of the attackers' tactics, techniques, and procedures (TTPs).

2. **Rapid decision-making:** Based on their analysis, the team made prompt decisions about the organization's security posture, allocating resources effectively, and implementing targeted response actions.
3. **Dynamic adaptation:** The team continuously adapted their response as added information emerged, adjusting their strategies and tactics to stay ahead of the attackers.
4. **Decentralized control:** The organization's emphasis on Implicit Guidance and Control allowed team members to make rapid decisions and take appropriate actions without waiting for explicit instructions from higher-ups, speeding up the response process.

Outcome

The financial services company's investment in Implicit Guidance and Control, coupled with their rapid OODA Loop strategic approach, resulted in the successful containment and mitigation of the cyber-attack. Notably, the incident was resolved significantly faster than similar incidents faced by other organizations in the industry, minimizing the potential damage to the company's reputation and customer trust.

Return on Investment

The organization's commitment to cultivating a culture of Implicit Guidance and Control yielded several tangible benefits:

1. **Faster incident resolution:** The rapid decision-making and adaptability fostered by Implicit Guidance and Control enabled the organization to respond more quickly to the attack, minimizing the potential damage and associated costs.
2. **Enhanced security posture:** The focus on continuous learning and adaptability strengthened the organization's overall security posture, making it more resilient to future threats.
3. **Improved team efficiency:** The decentralization of decision-making authority and the emphasis on diverse perspectives within the cybersecurity team led to more efficient and effective responses to incidents.

4. **Reduced risk exposure:** The organization's ability to anticipate and rapidly respond to emerging threats reduced the likelihood of significant financial losses, legal liabilities, and reputational damage.

In conclusion, the financial services company's investment in fostering Implicit Guidance and Control within its cybersecurity team demonstrated a clear return on investment by speeding up the OODA Loop and enabling a faster resolution of the incident, ultimately minimizing the potential damage and associated costs.

Case Study 7.2 – OODA in Third Party Risk Management

A detailed plan for incorporating the OODA Loop into a third-party risk management program can be broken down into the following steps:

A. **Identify third-party relationships:** Begin by compiling a comprehensive list of all third parties your organization works with, such as vendors, suppliers, contractors, and service providers. This will help you understand the scope of your third-party risk exposure.

1. **Observe:** Gather information on each third party's security practices, compliance with relevant regulations, and past performance. This can include requesting security documentation, reviewing audit reports, and researching any past incidents or breaches. Continuously monitor news and industry sources for any emerging threats, vulnerabilities, or trends that may affect your third parties.

2. **Orient:** Analyze the collected data to understand the risks associated with each third-party relationship. Identify key risk indicators, such as non-compliance with regulations, weak security controls, or a history of data breaches. Consider the potential impact of these risks on your organization, taking into account factors like the sensitivity of the data being shared,

the criticality of the services provided, and the potential reputational damage.

3. **Decide:** Based on your risk analysis, prioritize the identified risks and determine the most appropriate mitigation strategies. These strategies may include:
   i. Updating contractual requirements to include specific security controls or compliance obligations.
   ii. Requesting that third parties undergo regular security audits or assessments.
   iii. Implementing additional monitoring or reporting requirements.
   iv. Establishing contingency plans in case of third-party failure or service disruption.
   v. Identifying alternative suppliers or service providers to reduce dependence on high-risk third parties.

4. **Act:** Implement the chosen mitigation strategies and communicate your expectations to third parties. Ensure that your internal teams are aware of the third-party risk management program and their roles in supporting it. Monitor the progress of your third parties in meeting your requirements and addressing identified risks.

B. **Feedback and adaptation:** Continuously review and refine your third-party risk management program based on latest information, changing circumstances, or feedback from stakeholders. Update your risk assessments and mitigation strategies as needed to ensure that your program remains effective in addressing emerging risks and maintaining the security and resilience of your organization.

C. **Training and awareness:** Educate your employees and relevant stakeholders on the importance of third-party risk management and the role of the OODA Loop in the process. Provide training on how to identify, assess, and mitigate third-party risks effectively, and encourage a culture of continuous learning and improvement.

D. **Integration with other risk management frameworks:** Align your third-party risk management program with other risk management frameworks and methodologies used within your organization, such as the NIST Cybersecurity Framework or ISO/IEC 27000 series of standards. This will ensure a consistent and holistic approach to risk management across your organization.

E. **Metrics and KPIs:** Establish key performance indicators (KPIs) and metrics to measure the effectiveness of your third-party risk management program and its alignment with the OODA Loop principles. Regularly track and report on these metrics to demonstrate the value of the program and identify areas for improvement.

By following this plan and integrating the OODA Loop into your third-party risk management program, you can enhance your organization's ability to proactively identify, assess, and mitigate risks associated with third-party relationships, ensuring a more secure and resilient business environment.

In 2018, I worked in a team which authored a white paper for the Shared Assessments entitled "The OODA Loop: An Innovative Approach to Third-Party Risk Management" You can obtain the full paper at: **https://sharedassessments.org/register/?redirect-to=papers&product-type=papers&product-id=35993**. For the purposes of this book, I asked ChatGPT to provide a summary of the points in the paper and how they relate to the content of this book, which is as follows:

"The OODA Loop: An Innovative Approach to Third-Party Risk Management" provides an insightful exploration of how the OODA Loop can be applied to third-party risk management. The paper discusses the increasing importance of managing third-party risks due to growing dependence on external partners and vendors in today's interconnected business environment. The authors introduce the OODA Loop as a valuable framework for enhancing third-party risk management processes, stressing the significance of agility, adaptability, and strategic thinking.

The paper begins by outlining the four Phases of the OODA Loop: Observe, Orient, Decide, and Act. It highlights the iterative and cyclical nature of the loop, emphasizing that these Phases are interconnected and continuous. The authors then demonstrate how the OODA Loop can be integrated into third-party risk management processes, showcasing its potential for improving decision-making, risk identification, and response to emerging threats.

The paper discusses the application of the OODA Loop in the context of third-party risk management by breaking down each Phase and explaining its relevance. In the Observe Phase, organizations must gather information about their third-party partners, including their security posture, compliance status, and potential vulnerabilities. This Phase aligns with our previous discussions on the importance of threat intelligence and situational awareness in cybersecurity.

In the Orient Phase, organizations must analyze and interpret the collected data, considering their own goals, priorities, and risk tolerance. This Phase resonates with our conversations about the role of cultural traditions, genetic heritage, analysis, and synthesis in the OODA Loop. The author emphasizes the importance of context and understanding the broader threat landscape to make informed decisions.

The Decide Phase involves making decisions about the appropriate course of action based on the observations and orientations. The paper connects this Phase with our discussions on decision-making and the importance of Implicit Guidance and Control in the OODA Loop. It highlights the need for organizations to establish clear decision-making processes and criteria for evaluating third-party risks.

In the Act Phase, organizations implement their chosen course of action, whether it be accepting, mitigating, or transferring the identified risks. The paper echoes our discussions on the importance of speed, adaptability, and resilience in the OODA Loop. It also highlights the need for ongoing monitoring and feedback loops to ensure continuous improvement in third-party risk management.

Throughout the paper, the authors emphasize the importance of agility and adaptability in third-party risk management, reiterating the key themes we've discussed in relation to the OODA Loop and cybersecurity. The paper also touches upon the significance of communication, collaboration, and training in successfully applying the OODA Loop to third-party risk management, which aligns with our previous conversations.

In summary, this paper offers a valuable perspective on how the OODA Loop can be applied to third-party risk management, demonstrating its potential for enhancing decision-making, risk identification, and response to emerging threats. The authors' insights align with and complement the topics we've discussed in our conversations about the OODA Loop and cybersecurity, emphasizing the importance of agility, adaptability, strategic thinking, and continuous improvement."

CHAPTER 8: Implementing OODA in Cyber-security

"Machines don't fight wars. Terrain doesn't fight wars. Humans fight wars. You must get into the mind of humans. That's where the battles are won." - John Boyd

Blitzkrieg

When thinking about implementing the OODA Loop, I would be remiss if I didn't cover the concept of Blitzkrieg that Boyd studied. In the book "Certain to Win: The Strategy of John Boyd Applied to Business," Richards lays out the key attributes of the Blitzkrieg, which are:

- **Einheit:** Mutual trust, unity, and cohesion
- **Fingerspitzengefühl:** Intuitive feel, especially for complex and potentially chaotic situations
- **Auftragstaktik:** Mission, generally considered as a contract between superior and subordinate
- **Schwerpunkt:** Any concept that provides focus and direction to the operation

Let's dive into Blitzkrieg first. In the context of cybersecurity, incorporating Blitzkrieg-like tactics within the OODA Loop framework can help organizations become more agile and effective in responding to cyber threats. Blitzkrieg is a German term that translates to "lightning war." It refers to a military strategy that focuses on rapid, coordinated, and concentrated attacks to overwhelm and disorient the enemy, leading to their quick capitulation or disintegration. The concept of Blitzkrieg was developed and widely used by the German military during World War II, particularly in their campaigns against Poland and France.

The connection between the OODA Loop and Blitzkrieg lies in the emphasis on speed, decisiveness, and adaptability. In applying the OODA Loop decision-making process —Observe, Orient, Decide, and Act—the main objective is to move through these Phases more quickly than the adversary, thereby gaining an advantage in decision-making and action.

In the context of cybersecurity, the Blitzkrieg concept can be seen as a fast, decisive, and adaptive response to cyber threats. While Blitzkrieg is primarily a military strategy, its emphasis on speed, decisiveness, and adaptability aligns well with the principles of the OODA Loop. By rapidly identifying and responding to threats, organizations can effectively disrupt and neutralize cyber adversaries. Applying the principles of the OODA Loop in a Blitzkrieg-like manner, cybersecurity professionals can constantly stay ahead of attackers by quickly adapting to evolving threats and maintaining the initiative.

Einheit

Einheit is a German term that translates to "unity" or "cohesion." It refers to the harmonious collaboration and mutual understanding among members of a team or organization. Einheit emphasizes the importance of trust, shared values, and clear communication to enable effective decision-making and execution of actions in complex and dynamic environments.

Einheit is an essential component of the OODA Loop framework that enables a cohesive and unified approach to decision-making and action-taking, which is particularly important in fast-paced and complex environments like cybersecurity. Within this framework, Einheit plays a crucial role in enhancing the overall effectiveness and efficiency of the decision-making process. When team members work in unity and have a mutual understanding of goals, strategies, and tactics, they can better synchronize their observing, orienting, deciding, and acting efforts, and therefore can be more effective in response to various situations, including cybersecurity threats.

In the realm of cybersecurity, Einheit can be achieved by fostering a culture of collaboration, open communication, and shared responsibility. This unity among team members can lead to more effective threat detection, response, and mitigation, as well as better alignment with the organization's broader goals and objectives.

Fingerspitzengefühl

Fingerspitzengefühl is a German term that literally translates to "fingertip feeling." It is particularly valuable in the rapidly evolving world of cybersecurity, where threats and attack vectors are constantly changing. Fingerspitzengefühlrefers to the intuitive understanding or skill that someone has in a particular area or domain, often acquired through extensive experience and practice. Fingerspitzengefühl represents the ability to make quick, accurate judgments and decisions in complex and dynamic situations, often with limited information or time.

Fingerspitzengefühl plays a crucial role in enabling individuals or organizations to rapidly work through the four Phases of the OODA Loop in response to various situations, including cybersecurity threats. Developing this intuitive understanding can help professionals make better decisions within the OODA Loop and improve the overall effectiveness of their cybersecurity strategies. By developing a deep understanding of their domain, cybersecurity professionals can leverage their Fingerspitzengefühl to quickly assess threats, vulnerabilities, and potential consequences, as well as to determine the most effective course of action. Cultivating Fingerspitzengefühl often requires continuous learning, practice, and exposure to diverse situations to fine-tune one's intuition and improve decision-making capabilities.

Auftragstaktik

Auftragstaktik is a German term that translates to "mission command" or "mission tactics." It refers to a decentralized command and control approach where leaders provide a clear mission, objectives, and intent to their subordinates but allow them the flexibility and autonomy to determine the best way to achieve those objectives based on the situation on the ground.

Within the OODA Loop framework, Auftragstaktik plays the essential role of promoting agility, adaptability, and rapid decision-making in complex and dynamic environments. This is an important concept within the OODA Loop framework as it emphasizes decentralized decision-making and flexibility in execution, allowing organizations to better adapt and respond to rapidly evolving situations, such as those frequently encountered in the field of cybersecurity. By empowering individuals and teams to make decisions and take actions based on their understanding of the situation, organizations can respond more effectively to changing circumstances, including cybersecurity threats.

In cybersecurity, Auftragstaktik can be applied by providing clear goals, objectives, and intent to cybersecurity teams while allowing them the flexibility to choose the most appropriate tactics, techniques, and tools to achieve those objectives. This decentralized approach can lead to more effective threat detection, response, and mitigation and—importantly, foster a culture of innovation and adaptability.

Schwerpunkt

Schwerpunkt is a German term that translates to "center of gravity" or "focal point." The concept of Schwerpunkt was originally developed in the context of military strategy, particularly in the works of Prussian general and military theorist Carl von Clausewitz. In military strategy, Schwerpunkt represents the key point of an operation or the primary objective that, if achieved, can lead to victory or bring about the enemy's collapse.

In the context of the OODA Loop, Schwerpunkt refers to the focal point of an organization's decision-making and action-taking process. This focal point is often the main objective or goal that guides an organization's efforts in responding to various situations, including cybersecurity threats. In cybersecurity, the Schwerpunkt could be the protection of sensitive data, ensuring system uptime, or maintaining the integrity of the organization's digital assets. By identifying and concentrating on the Schwerpunkt and focusing resources and efforts on it, cybersecurity professionals can make more informed decisions within the OODA Loop framework, prioritize their actions, and make more effective decisions to achieve the most significant impact.

OODA Adoption Challenges

"The difficulty lies not so much in developing new ideas as in escaping from old ones." - John Maynard Keynes

While the OODA Loop is a natural decision-making process, there are several reasons why it might not be as widely adopted in business and cybersecurity as one might expect:

1. **Lack of awareness:** Many people may not be familiar with the OODA Loop concept or its potential applications in business and cybersecurity. Without an understanding of the framework and its benefits, people may not recognize the value of implementing it in their professional environments.

2. **Complexity of situations:** Business and cybersecurity situations can be extraordinarily complex, involving multiple stakeholders, varying objectives, and rapidly changing circumstances. This complexity can make it challenging for individuals or organizations to apply the OODA Loop effectively, as it requires continuous information gathering, analysis, and adaptation.

3. **Resistance to change:** Organizations often have established decision-making processes, and employees may be resistant to change or reluctant to adopt new methods. This resistance can be particularly strong if the current processes appear to be working well or if there is a perception that the OODA Loop is too time-consuming or resource intensive.

4. **Cognitive biases and heuristics:** People naturally rely on cognitive shortcuts, or heuristics, to simplify decision-making. These shortcuts can lead to biases that may hinder the effective application of the OODA Loop. For example, individuals may overvalue their own experiences or opinions, underestimate the potential impact of new information, or be overly influenced by group dynamics.

5. **Inadequate training and education:** Successfully applying the OODA Loop in business and cybersecurity requires training and education to develop the skills necessary for effective information gathering, analysis, decision-making,

and action. Organizations may not invest in the necessary training or may not prioritize developing these skills in their employees.

Despite these challenges, organizations that make the effort to employ the OODA Loop will find that it has the potential to significantly improve decision-making and responsiveness in both business and cybersecurity contexts. By raising awareness of the framework, investing in training and education, and addressing potential barriers to adoption, organizations can better leverage the OODA Loop to enhance their strategic capabilities and resilience in the face of rapidly evolving threats and opportunities.

Challenges and Barriers to Implementing the OODA Loop in Cybersecurity Practices and Strategies to Overcome Them

This discussion will explore various challenges and barriers organizations may encounter when implementing the OODA Loop in their cybersecurity practices and propose strategies to overcome these obstacles. These challenges can hinder the effective adoption and application of the OODA Loop methodology in addressing cybersecurity threats.

1. **Resistance to change:** As noted, one of the most significant challenges organizations may face when implementing the OODA Loop in their cybersecurity practices is resistance to change. Employees may be hesitant to adopt a new methodology or adapt to new processes, particularly if they are already accustomed to existing practices. To overcome this resistance, leaders should communicate the benefits and rationale behind adopting the OODA Loop, provide ongoing support and training, and actively involve employees in the implementation process to foster a sense of ownership and commitment.

2. **Lack of understanding and awareness:** Another challenge organizations may encounter is a lack of understanding and awareness about the OODA Loop and its potential benefits.

This can hinder the effective adoption and application of the OODA Loop methodology. To address this challenge, leaders should invest in education and training programs to enhance employees' understanding of the OODA Loop, its principles, and its potential benefits for the organization's cybersecurity posture. Methods may involve conducting workshops, seminars, or online training sessions to ensure that employees are well-equipped to apply the OODA Loop in their day-to-day activities.

3. **Insufficient resources:** Implementing the OODA Loop in an organization's cybersecurity practices may require additional resources, including personnel, technology, and funding. Organizations may struggle to allocate the necessary resources to support the OODA Loop implementation, particularly if they are already facing budget constraints or competing priorities. To overcome this challenge, leaders should prioritize the OODA Loop implementation in their strategic planning, allocate resources accordingly, and seek out opportunities to leverage existing resources, partnerships, or grants to support the initiative.

4. **Inadequate information sharing and collaboration:** The OODA Loop relies heavily on information sharing and collaboration, both internally and externally. Organizations may face challenges in establishing effective channels for sharing information and insights, as well as fostering a culture of trust and transparency. To address this challenge, leaders should prioritize the development of robust communication and collaboration mechanisms, including secure platforms for sharing threat intelligence, regular meetings or briefings to discuss emerging threats and vulnerabilities, and partnerships with external organizations to facilitate the sharing of best practices and threat intelligence.

5. **Organizational silos:** Organizational silos can be a significant barrier to the effective implementation of the OODA Loop in cybersecurity practices. Silos can impede the flow of

information and hinder collaboration between teams, which is crucial to the success of the OODA Loop methodology. To overcome this obstacle, leaders should actively work to break down silos and foster cross-functional collaboration, including the establishment of interdisciplinary teams, regular cross-functional meetings, and the integration of cybersecurity responsibilities across divergent functions and departments.

6. **Legal and regulatory constraints:** Organizations may also face legal and regulatory constraints that hinder the effective application of the OODA Loop in their cybersecurity practices. Compliance with data protection laws and regulations—such as GDPR or CCPA—is crucial, and security teams should ensure that their actions align with these requirements. To address this challenge, organizations should work closely with legal and regulatory experts to ensure that their OODA Loop implementation is compliant with all applicable laws and regulations and develop processes and procedures to regularly review and update their practices in response to evolving legal and regulatory requirements.

7. **Inadequate performance measurement and feedback:** For the OODA Loop to be effective in a cybersecurity context, organizations must be able to measure its performance and gather feedback to inform continuous improvement efforts. This can be challenging, particularly when it comes to identifying the appropriate key performance indicators (KPIs) and metrics that capture the impact of the OODA Loop on the organization's cybersecurity posture. To overcome this challenge, organizations should invest in the development of a robust performance measurement framework that includes clear and relevant KPIs and metrics, as well as mechanisms for gathering and analyzing feedback from various stakeholders.

8. **Balancing speed and accuracy:** The OODA Loop emphasizes the need for rapid decision-making and action in response to cybersecurity threats. However, organizations must also

ensure that the decisions made, and actions taken are accurate and well-informed. Achieving this can be challenging, particularly when dealing with complex and evolving threats. To address this challenge, organizations should invest in advanced analytical tools and technologies that can support rapid yet informed decision-making, as well as training and education initiatives that enhance the decision-making capabilities of their cybersecurity teams.

9. **Maintaining situational awareness:** The OODA Loop relies heavily on maintaining situational awareness to inform decision-making and action. Organizations may struggle to maintain a comprehensive and up-to-date understanding of the threat landscape, particularly as the number and complexity of cybersecurity threats continue to grow. To overcome this challenge, organizations should invest in threat intelligence platforms and services, as well as training and education initiatives, to ensure that their security teams are well-equipped to maintain situational awareness and respond effectively to emerging threats and vulnerabilities.

10. **Managing complexity and uncertainty:** The implementation of the OODA Loop in cybersecurity practices can be complex and uncertain, particularly as organizations navigate a rapidly evolving threat landscape and an increasingly interconnected digital ecosystem. To address this challenge, organizations should embrace a culture of adaptability and resilience, regularly reviewing and updating their security policies, processes, and technologies to ensure they remain effective in the face of new challenges. This may involve conducting periodic risk assessments, scenario planning exercises, and red teaming exercises to identify potential vulnerabilities and develop strategies for addressing them.

By addressing these challenges and barriers, organizations can successfully implement the OODA Loop in their cybersecurity practices and improve their ability to detect, analyze, and respond to cybersecurity threats. With its focus on continuous improvement, collaboration, and adaptability, the OODA Loop can serve as a powerful tool for enhancing an organization's cybersecurity posture and resilience in the face of an ever-evolving threat landscape.

The role of leadership and organizational culture in the successful implementation of the OODA Loop

In this discussion, we will delve into various aspects of leadership and organizational culture that contribute to the successful application of the OODA Loop in cybersecurity programs.

First and foremost, strong leadership and organizational culture play a crucial role in the successful implementation of the OODA Loop in cybersecurity programs. Leaders can effectively leverage the OODA Loop to strengthen their organization's cybersecurity posture by fostering a culture of continuous improvement, empowering their teams to make informed decisions, and encouraging collaboration and knowledge sharing throughout the organization.

Leaders must understand the importance of the OODA Loop and its potential benefits, as well as demonstrate a commitment to integrating the OODA Loop into the organization's cybersecurity strategy. This may involve setting clear expectations and goals, allocating resources, and providing the necessary support and guidance to help teams adapt to the OODA Loop methodology.

Furthermore, leaders play a pivotal role in fostering a culture of continuous improvement within the organization. By promoting a mindset of learning, adaptation, and innovation, leaders can encourage their teams to continually refine and enhance the OODA Loop process. This may involve regularly reviewing and updating security policies, processes, and technologies to ensure they remain effective in the face of new challenges. Additionally, leaders should be open to receiving feedback from their teams and willing to adjust their strategies and tactics as needed.

Empowerment is another critical aspect of leadership in the context of the OODA Loop. Leaders should empower their teams to make informed decisions, providing them with the necessary autonomy and authority to act decisively in response to cybersecurity threats. This may involve delegating decision-making authority to lower levels of the organization, ensuring that decisions can be made quickly and effectively in the face of rapidly evolving threats. By fostering a culture of empowerment, leaders can accelerate the OODA Loop process and improve the organization's overall cybersecurity posture.

Collaboration and knowledge sharing are also essential components in successfully implementing the OODA Loop. Leaders should encourage open communication and collaboration, both within and between teams, to enhance the collective understanding of the threat landscape and improve the organization's overall cybersecurity posture. This may involve establishing clear channels for sharing information and insights, as well as promoting a culture of trust and transparency. Additionally, leaders should foster partnerships with external organizations, such as industry groups and government agencies, to facilitate the sharing of threat intelligence and best practices.

Organizational culture plays a significant role in the successful implementation of the OODA Loop in cybersecurity programs. A culture that values learning, adaptation, collaboration, and empowerment will more likely embrace the OODA Loop methodology and effectively apply it to address cybersecurity challenges. As such, leaders must actively work to shape and influence the organization's culture, demonstrating through their actions and behaviors the importance of the OODA Loop and its potential benefits.

As the cyber threat landscape evolves, the need for agile, adaptive, and resilient cybersecurity programs becomes increasingly important. By embracing the OODA Loop methodology and cultivating a supportive organizational culture, leaders can equip their organizations to navigate the challenges of the modern cybersecurity environment and emerge stronger and more secure.

CHAPTER 9: Listening to the Experts

"Standing on the shoulders of giants" - Isaac Newton

As an author who has spent over a decade studying and applying the OODA Loop in the field of cybersecurity, I have been deeply influenced by several seminal works that have shaped my understanding of this powerful decision-making framework. Articles, including Boyd's "Patterns of Conflict," "Destruction and Creation," "The Essence of Winning and Losing," as well as the insightful piece on the Art of Manliness website, have each been instrumental in the development of my methodology for using the OODA Loop in cybersecurity.

These sources emphasize the importance of the OODA Loop as a means of achieving a competitive advantage in complex and dynamic environments, such as the ever-evolving cyber landscape. They also highlight the critical role of human factors, adaptability, and continuous learning in the successful application of the OODA Loop. Drawing upon the insights provided by these influential works, I have identified seven key themes that serve as the cornerstone of my methodology for using the OODA Loop in cybersecurity. Each is essential to the overall understanding and effective implementation of the OODA Loop in cybersecurity:

- Adaptability and Agility
- Decision-Making Speed
- Continuous Learning and Improvement
- Holistic Approach
- Importance of Human Factors
- Information as a Key Resource
- Emphasis on Competitive Advantage

By exploring and applying these themes, organizations can enhance their security posture, stay ahead of emerging threats, and ultimately, ensure their long-term resilience in the face of an increasingly complex and hostile cyber environment. I will expand on those themes later, but first let's dive into the analysis of the documents.

Common Themes from the Experts

Several common themes emerge when analyzing key background documents, which are applicable to both the OODA Loop and cybersecurity:

1. **Adaptability and Agility**: In the context of cybersecurity, adaptability and agility refer to the ability of an organization to rapidly adjust its strategies, processes, and technologies in response to evolving threats and changing circumstances. This may involve adopting new security tools, revising security policies, or reallocating resources to address emerging risks. Organizations that are adaptable and agile can better anticipate and respond to cyber threats, reducing the likelihood of successful attacks and minimizing the impact of security incidents.

2. **Decision-Making Speed**: Rapid decision-making is a critical component of the OODA Loop and is particularly important in cybersecurity, where threats can emerge and evolve quickly. Organizations need to develop processes and systems that enable them to make informed decisions promptly, such as real-time monitoring and analysis of security events, automated response capabilities, and streamlined incident response processes. By accelerating decision-making, organizations can more effectively defend against cyber threats and reduce the potential for severe damage.

3. **Continuous Learning and Improvement:** In the dynamic world of cybersecurity, continuous learning and improvement are essential for staying ahead of attackers and adapting to the ever-changing threat landscape. This involves regularly reviewing and updating security policies and procedures, incorporating lessons learned from past incidents, and staying informed about the latest trends and best practices in cybersecurity. Organizations should also foster a culture of learning and improvement, encouraging employees to seek ongoing professional development opportunities and share their knowledge with colleagues.

4. **Holistic Approach**: A holistic approach to cybersecurity involves considering all aspects of security and ensuring that they are integrated and aligned with the organization's overall goals and objectives. This includes not only technical measures such as firewalls, intrusion detection systems, and encryption but also organizational elements like security policies, user training, and incident response plans. By taking a holistic approach, organizations can more effectively identify and address potential vulnerabilities and ensure a comprehensive and cohesive security strategy.

5. **Importance of Human Factors**: Human factors play a crucial role in cybersecurity, as people are often the weakest link in an organization's security posture. This includes both unintentional actions, such as falling victim to phishing attacks or inadvertently disclosing sensitive information, and malicious insider threats. To address human factors, organizations must invest in employee training and awareness programs, create and enforce security policies that are easy to understand and follow, and foster a security-conscious culture.

6. **Information as a Key Resource**: In the OODA Loop and cybersecurity, information is a critical resource that informs decision-making and drives action. Organizations must prioritize the collection, analysis, and sharing of relevant information, including threat intelligence, vulnerability data, and security event logs. Effective information management helps organizations make informed decisions about their security posture, identify and respond to emerging threats, and maintain situational awareness in a complex and evolving cyber environment.

7. **Emphasis on Competitive Advantage**: Gaining a competitive advantage in cybersecurity involves staying one step ahead of attackers by anticipating their moves, proactively defending against threats, and adapting strategies based on the evolving threat landscape. Organizations can achieve a competitive advantage by investing in advanced security technologies, leveraging threat intelligence, and

continuously improving their security processes and capabilities. By maintaining a competitive edge, organizations can better protect their assets, minimize the risk of security breaches, and ensure their long-term resilience in the face of cyber threats.

By understanding and applying these common themes, organizations can leverage the OODA Loop as a valuable framework for enhancing their cybersecurity strategies and improving their overall security and resilience in the face of an ever-changing cyber threat environment.

Destruction and Creation
Commentary on Boyd's Paper and Its Relevance to OODA and Cybersecurity

In this commentary, we will analyze the key concepts presented in "Destruction & Creation" and explore their relevance to the OODA Loop and cybersecurity. "Destruction & Creation" was written by Colonel John Boyd in 1976. This paper offers an in-depth look at the process of decision-making and the importance of adaptability in the face of uncertainty. The paper lays the foundation for Boyd's later development of the OODA Loop, a decision-making framework widely used in military strategy and other domains, including cybersecurity. By understanding the key concepts presented in the paper, such as the process of destruction and creation, conceptual spirals, inductive and deductive reasoning, analogy, unfettered imagination, and rapid adaptation, cybersecurity professionals can better apply the principles of the OODA Loop to their security strategies and enhance their organization's ability to respond to evolving threats.

The Process of Destruction and Creation

Boyd's paper begins by describing the process of destruction and creation, which involves breaking down existing concepts and ideas and synthesizing new ones to adapt to changing circumstances. Boyd argues that this process is essential for decision-makers to cope with uncertainty and devise effective strategies in a constantly evolving environment.

In the context of cybersecurity, the process of destruction and creation is highly relevant, as organizations must continuously adapt their security strategies to address rapidly changing emerging threats and vulnerabilities. This environment requires security teams to constantly question and reassess their existing understanding of the threat landscape, their security posture, and their strategies; and to develop innovative approaches based on the changing dynamics of the cyber environment.

Conceptual Spirals and the Synthesis of New Concepts

Within the larger topic, Boyd introduces the idea of conceptual spirals, which involves the continuous destruction and creation of concepts to adapt to an ever-changing environment. He emphasizes the importance of being able to rapidly shift between different perspectives and mental models to make sense of complex situations and make informed decisions.

In cybersecurity, conceptual spirals can be applied to the continuous process of evaluating and updating security strategies based on the evolving threat landscape. Security teams must be able to rapidly assess the effectiveness of their current strategies, identify areas for improvement, and develop innovative approaches to address emerging threats and vulnerabilities. This requires an elevated level of mental agility and the ability to synthesize new concepts from diverse sources of information.

Inductive and Deductive Reasoning

Boyd discusses the roles of inductive and deductive reasoning within the process of destruction and creation. Both types of reasoning are essential for the process of destruction and creation, as they enable decision-makers to create and validate new concepts in response to changing circumstances.

- Inductive reasoning involves generating new concepts based on specific observations, while
- Deductive reasoning involves testing the validity of these concepts against a broader set of observations.

In the context of the OODA Loop and cybersecurity, both inductive and deductive reasoning play critical roles in the Orientation Phase, where security teams analyze observed data and synthesize new understanding of the cyber threat landscape. By employing both types of reasoning, security teams can develop a comprehensive understanding of the threats they face and devise effective strategies to address them.

The Role of Analogy in Decision-Making

Boyd highlights the importance of analogy in decision-making, as it allows individuals to draw connections between seemingly unrelated concepts and apply their understanding of one domain to another. This ability to make connections and transfer knowledge is essential for decision-makers operating in complex and uncertain environments.

In the realm of cybersecurity, the use of analogy can help security teams better understand the motives and tactics of adversaries, as well as the potential impact of threats on their organization's operations and assets. By drawing on their understanding of similar situations or historical incidents, security teams can develop more effective strategies and anticipate potential challenges.

The Importance of Unfettered Imagination

Boyd emphasizes the importance of unfettered imagination in the process of destruction and creation, arguing that decision-makers must be willing to challenge existing paradigms and think beyond conventional wisdom to adapt to changing circumstances. This requires a willingness to embrace uncertainty and take risks in the pursuit of innovative solutions.

In cybersecurity, fostering a culture of unfettered imagination is crucial for staying ahead of rapidly evolving threats. Security teams must be willing to explore unconventional approaches and experiment with innovative technologies to enhance their organization's security posture. This may involve embracing emerging technologies such as artificial intelligence, machine learning, and blockchain, or exploring new collaboration models with industry partners and government agencies.

The Need for Rapid Adaptation

Boyd's paper underscores the importance of rapid adaptation in decision-making, as it enables decision-makers to stay ahead of their adversaries and respond effectively to changing circumstances. This emphasis on speed and agility is a core principle of the OODA Loop and is particularly relevant in the cybersecurity domain.

To enhance their ability to adapt rapidly, organizations can:

- Implement advanced monitoring and analytics tools to automate the collection and analysis of large volumes of data, enabling security teams to quickly identify and respond to emerging threats.
- Streamline decision-making processes and ensure clear communication channels among team members, allowing for rapid coordination and execution of security strategies.
- Foster a culture of continuous learning and improvement, ensuring that security teams stay up to date with the latest threats, vulnerabilities, and best practices.

An understanding of these key concepts introduced in Colonel John Boyd's "Destruction & Creation" provides valuable insights into the process of decision-making and the importance of adaptability in uncertain and complex environments such as cyber.

Chapter Sources

Boyd, J. R. (1976). Destruction & Creation. Retrieved from https://upload.wikimedia.org/wikipedia/commons/a/a6/Destructio n_%26_Creation.pdf

Essence of Winning and Losing
Commentary on Boyd's Presentation and Its Relevance to OODA and Cybersecurity

This commentary will analyze the key concepts presented in the presentation "The Essence of Winning and Losing" created by Colonel John Boyd in 1995 and discuss their relevance to the OODA Loop concepts already explored in the context of cybersecurity. This presentation provides valuable insights into the OODA Loop and its applications in decision-making and competitive environments.

The Importance of Time and Tempo

Boyd's presentation emphasizes the importance of time and tempo in decision-making, particularly in competitive environments. He argues that by operating at a faster tempo than one's adversaries, it is possible to disrupt their decision-making processes and gain a decisive advantage. This concept is central to the OODA Loop, which stresses the need for rapid and continuous cycles of observation, orientation, decision, and action.

In cybersecurity, the importance of time and tempo cannot be overstated. By acting faster than threat actors, organizations can identify, respond to, and mitigate threats before they can cause considerable damage. To achieve a faster tempo, organizations can implement advanced monitoring and analytics tools, streamline decision-making processes, and foster a culture of continuous learning and improvement.

The Role of Implicit Guidance and Control

In this presentation, Boyd introduces the concept of Implicit Guidance and Control, which involves using intuition, experience, and mental models to guide decision-making and action, rather than relying solely on explicit instructions or plans. Implicit Guidance and Control can enable decision-makers to respond more quickly and effectively to rapidly changing circumstances, as they can adapt their actions based on their understanding of the situation rather than waiting for new instructions or information specific to the changed environment.

In the context of cybersecurity, Implicit Guidance and Control can be applied by empowering security teams to make decisions and act based on their experience, intuition, and understanding of the threat landscape. Using Implicit Guidance and Control requires organizations to invest in the development of their security teams' expertise, as well as create an environment that encourages innovation, experimentation, and collaboration.

The Significance of Interaction with the Environment

Boyd's presentation highlights the significance of interacting with the environment and the importance of feedback in decision-making. In the OODA Loop, the process of observing, orienting, deciding, and acting is an iterative cycle that continuously adapts based on feedback from the environment. This continuous interaction allows decision-makers to refine their understanding of the situation, develop more effective strategies, and anticipate potential challenges.

In cybersecurity, organizations must maintain a continuous interaction with the cyber environment to stay informed of emerging threats, vulnerabilities, and best practices. This involves monitoring the environment, collecting and analyzing data, and collaborating with industry peers, government agencies, and other stakeholders. By continuously interacting with the environment, organizations can better anticipate and respond to evolving threats.

The Snowmobile Analogy

Boyd uses the snowmobile analogy to illustrate the process of destruction and creation in decision-making. The snowmobile is constructed by combining parts from a tank, a bicycle, a boat, and a jet engine, representing the synthesis of disparate concepts to create a new and innovative solution. This analogy highlights the importance of mental agility and the ability to synthesize innovative ideas from diverse sources of information in the face of uncertainty and complexity.

In the realm of cybersecurity, the snowmobile analogy can serve as a reminder of the importance of innovation and the need to think beyond conventional wisdom to address emerging threats and vulnerabilities. By fostering a culture of experimentation and encouraging security teams to explore unconventional approaches, organizations can enhance their ability to adapt to the rapidly evolving cyber threat landscape.

Chapter Sources

Boyd, J. R. (1995). The Essence of Winning and Losing. Retrieved from **https://fasttransients.files.wordpress.com/2020/03/boydsoodalo opnecesse-1.pdf**

Patterns of Conflict
Commentary on Boyd's Briefing and Its Relevance to OODA and Cybersecurity

This commentary will analyze the key concepts presented in the comprehensive briefing "Patterns of Conflict" created by Colonel John Boyd in 1986 and discuss their relevance to the OODA Loop concepts already explored in the context of cybersecurity. This briefing provides an in-depth analysis of conflict and warfare throughout history and offers valuable insights into the OODA Loop and its applications in decision-making and conflict resolution.

The Nature of Conflict and Competition

Boyd's briefing examines the nature of conflict and competition, focusing on the common patterns and underlying principles that can be observed across different historical examples. He identifies several key factors that contribute to the outcome of conflicts, such as the ability to adapt to changing circumstances, the capacity to disrupt an adversary's decision-making processes, and the importance of maintaining a cohesive and focused force.

In the context of cybersecurity, these principles can be applied to understand the dynamics of the cyber threat landscape and develop more effective security strategies. Organizations must be able to adapt to the constantly evolving nature of cyber threats, disrupt the efforts of adversaries, and maintain a cohesive and focused security team to effectively protect their digital assets and infrastructure.

The Concept of Cheng and Ch'i

Boyd introduces the concept of Cheng and Ch'i in this briefing. These combined concepts are derived from ancient Chinese military strategy, which emphasizes the need to balance direct (Cheng) and indirect (Ch'i) approaches to achieve success in conflict. This can be applied to the OODA Loop, where organizations must strike a balance between direct, visible actions (such as deploying security tools and technologies) and indirect, less visible actions (such as gathering intelligence on adversaries and enhancing internal security processes) to enhance their security posture. This may involve implementing advanced security technologies and conducting visible security audits, while also engaging in threat intelligence gathering, collaborating with industry partners, and fostering a culture of security awareness within the organization.

The Principle of Agility

Boyd emphasizes the importance of agility in conflict and decision-making as a critical factor in achieving success, arguing that the ability to rapidly adapt to changing circumstances and outmaneuver adversaries. This principle is central to the OODA Loop, which highlights the need for continuous cycles of observation, orientation, decision, and action to maintain a competitive advantage.

In the realm of cybersecurity, agility is crucial for organizations to stay ahead of evolving threats and vulnerabilities. This requires security teams to rapidly assess their security posture, identify areas for improvement, and develop innovative approaches to address emerging threats. By fostering a culture of continuous learning and improvement, organizations can enhance their agility and resilience in the face of cyber threats.

The Role of Moral Conflict

Boyd's briefing highlights the importance of moral conflict in determining the outcome of confrontations, emphasizing the need to undermine an adversary's morale and cohesion while maintaining one's own. This aspect of conflict is relevant to the OODA Loop and cybersecurity, as organizations must strive to maintain the trust and confidence of their stakeholders (including employees, customers, and partners) while seeking to erode the morale and cohesion of threat actors.

In cybersecurity, organizations can apply the principle of moral conflict by demonstrating a strong commitment to security, fostering a culture of transparency and accountability, and actively engaging with stakeholders to address concerns and build trust. Simultaneously, cybersecurity professionals can seek to undermine the morale and cohesion of adversaries by sharing threat intelligence, collaborating with law enforcement agencies, and actively disrupting cybercriminal operations.

Colonel John Boyd's "Patterns of Conflict" briefing offers valuable insights into the nature of conflict, competition, and decision-making, which can be applied to the OODA Loop and cybersecurity. By understanding the principles and concepts presented in the briefing, such as the nature of conflict and competition, the concept of Cheng and Ch'i, the principle of agility, and the role of moral conflict, cybersecurity professionals can better apply the principles of the OODA Loop to their security strategies and enhance their organization's ability to respond to evolving threats.

The Concept of Mission Command

Boyd introduces the concept of mission command, which involves decentralizing decision-making authority to lower levels of command while maintaining a clear overarching intent. Mission Command is the overarching support for successful Implicit Guidance and Control as discussed previously. This approach allows for greater flexibility and adaptability in the face of rapidly changing circumstances, as individual units can respond to local conditions and make decisions without waiting for instructions from higher levels of command.

In the context of cybersecurity, organizations can apply the concept of mission command by empowering their security teams to make decisions and act based on their understanding of the threat landscape and the organization's overall security objectives. This requires organizations to invest in the development of their security teams' expertise, provide clear guidance on overarching objectives, and create an environment that encourages collaboration and communication between various levels of the organization.

The Power of Initiative

Boyd emphasizes the importance of seizing and maintaining the initiative in conflict, as it allows decision-makers to shape the unfolding situation and force adversaries to react to their actions, rather than the other way around. This principle is closely related to the OODA Loop, as organizations that can cycle through the loop faster than their adversaries can maintain the initiative and dictate the course of the conflict.

In cybersecurity, organizations must strive to maintain the initiative by staying ahead of emerging threats and proactively implementing measures to protect their digital assets and infrastructure. This may involve continuously monitoring the threat landscape, investing in innovative security technologies, and engaging in active threat hunting to identify and disrupt potential attacks before they can cause harm.

Colonel John Boyd's "Patterns of Conflict" provides a comprehensive analysis of conflict and warfare throughout history, offering valuable insights into the principles and concepts that underpin the OODA Loop and its applications in decision-making and conflict resolution. By understanding and applying these principles and concepts, cybersecurity professionals can enhance their organization's ability to respond to evolving cyber threats and maintain a competitive advantage in the ever-changing cyber threat landscape.

Chapter Sources

Boyd, J. R. (1986). Patterns of Conflict. Retrieved from https://dnipogo.org/boyd/pdf/poc.pdf

The OODA Loop and Cybersecurity
Insights from The Art of Manliness

The Art of Manliness, a popular men's lifestyle blog, has published an article on the OODA Loop, a decision-making framework developed by Colonel John Boyd. The article provides an accessible overview of the OODA Loop and its applications in various aspects of life, including personal development, business, and even sports. This summary will explore the key insights from the article and discuss their relevance to the OODA Loop concepts already presented in the context of cybersecurity.

Applying the Art of Manliness' Insights to Cybersecurity

The Art of Manliness' article emphasizes the importance of speed and adaptability in executing the OODA Loop, as well as the need to maintain situational awareness, enhance mental agility, and develop mental models. These insights can be applied to the cybersecurity domain to improve the effectiveness of security strategies and enhance an organization's ability to respond to evolving threats.

Speed and Adaptability

In the context of cybersecurity, speed and adaptability are crucial for staying ahead of adversaries who continuously develop new tactics and techniques to exploit vulnerabilities. Security teams must quickly observe, orient, decide, and act to mitigate threats before they can cause considerable damage.

To enhance speed and adaptability, organizations can:

- Invest in advanced analytics and machine learning technologies to automate the analysis of large volumes of data.
- Develop streamlined decision-making processes and ensure clear communication channels among team members.
- Foster a culture of continuous learning and improvement, ensuring that security teams stay up to date with the latest threats and best practices.

Situational Awareness

Situational awareness—or the ability to understand the environment and the implications of observed events—is crucial for effective decision-making in cybersecurity. To maintain situational awareness, security teams must continually collect and analyze data, incorporate latest information, and adapt their understanding of the threat landscape.

Organizations can enhance situational awareness by:

- Establishing robust threat intelligence programs to gather information on emerging threats and vulnerabilities.
- Sharing information and collaborating with industry peers, government agencies, and other stakeholders.
- Encouraging security teams to stay informed of the latest research, news, and developments in cybersecurity.

Mental Agility

Mental agility, or the ability to rapidly shift between different mental models and perspectives, can help security teams better anticipate and respond to evolving threats. In the context of cybersecurity, mental agility involves understanding the motives and tactics of adversaries, as well as the potential impact of threats on the organization's operations and assets.

To develop mental agility, organizations can:

- Provide regular training and education opportunities for security teams to enhance their understanding of the threat landscape and develop their analytical skills.
- Encourage security teams to engage in scenario planning and threat modeling exercises to explore potential threats and their implications from various perspectives.
- Foster a culture of innovation and experimentation, allowing security teams to test and refine new strategies and approaches.

Mental Models

Mental models are simplified representations of the world that help individuals make sense of complex situations and predict the outcomes of their actions. In cybersecurity, mental models can be used to understand the behaviors and tactics of adversaries, the potential vulnerabilities within an organization's systems, and the potential consequences of various security measures.

To develop and leverage mental models effectively, organizations can:

- Encourage security teams to create threat profiles of various adversaries, including their motives, capabilities, and preferred tactics.
- Utilize frameworks such as <u>MITRE ATT&CK®</u> to map out potential attack vectors and the techniques commonly used by adversaries.
- Regularly review and update mental models to account for the changing threat landscape and the organization's evolving security posture.

Boyd's Law of Iteration

The Art of Manliness' article introduces Boyd's Law of Iteration, which states that the success of a decision-making process depends more on the speed of iteration than the accuracy of each individual decision. In other words, *it is more important to rapidly cycle through the OODA Loop and adapt to changing circumstances than to strive for perfect decision-making.*

In the context of cybersecurity, this principle underscores the importance of agility and adaptability in responding to threats. Organizations can benefit from the following strategies:

- **Embrace a proactive approach** to security by continuously monitoring and analyzing the environment, updating their understanding of the threat landscape, and adjusting their strategies accordingly.
- **Encourage security teams to iterate rapidly** through the OODA Loop by streamlining decision-making processes, automating routine tasks, and focusing on high-impact actions.
- **Learn from both successes and failures** by conducting post-incident reviews and incorporating lessons learned into future strategies and decision-making processes.

The Art of Manliness' insights on the OODA Loop can provide valuable guidance for organizations seeking to improve their cybersecurity posture. By emphasizing speed and adaptability, maintaining situational awareness, enhancing mental agility, developing mental models, and applying Boyd's Law of Iteration, organizations can better anticipate and respond to evolving threats in the complex and dynamic world of cybersecurity.

Chapter Sources

McKay, B., & McKay, K. (2014). The OODA Loop: How to Turn Uncertainty into Opportunity in a Chaotic World. The Art of Manliness. Retrieved from **https://www.artofmanliness.com/character/behavior/ooda-loop/**

CHAPTER 10: Expanding on OODA

"Innovation distinguishes between a leader and a follower." - Steve Jobs

Integrating the OODA Loop with Other Cybersecurity Frameworks and Methodologies for Comprehensive Risk Management

The OODA Loop, with its cyclical Phases of Observe, Orient, Decide, and Act, offers a valuable approach to managing cybersecurity risks. However, to fully harness its potential, it is important to integrate the OODA Loop with other well-established cybersecurity frameworks and methodologies. By doing so, organizations can create a comprehensive approach to cybersecurity risk management that addresses the ever-evolving threat landscape.

Three prominent examples of cybersecurity frameworks and methodologies that can be integrated with the OODA Loop to enhance an organization's cybersecurity posture are the NIST Cybersecurity Framework, MITRE ATT&CK, and the ISO/IEC 27000 series of standards.

- **The NIST Cybersecurity Framework**, developed by the United States National Institute of Standards and Technology, is a voluntary framework that provides organizations with guidelines for managing and reducing cybersecurity risks. This Framework is built around five core functions: Identify, Protect, Detect, Respond, and Recover. These functions can be mapped to the OODA Loop, enhancing its effectiveness. For instance, the Identify and Detect functions align with the Observe Phase, while the Protect function relates to the Orient Phase. The Respond function corresponds with the Decide and Act Phases, and the Recover function can be considered an extension of the Act Phase, emphasizing the importance of learning and improvement.

- **MITRE ATT&CK (Adversarial Tactics, Techniques, and Common Knowledge)** is a knowledge base and framework that captures information about the tactics, techniques, and procedures (TTPs) employed by cyber adversaries. It provides a structured approach to understanding and analyzing the behavior of cyber threat actors. By integrating MITRE ATT&CK into the OODA Loop, organizations can enhance their ability to Observe and Orient by having a better understanding of the TTPs used by adversaries. This improved understanding can lead to more informed decisions (Decide Phase) and more effective responses (Act Phase) to cyber threats.

- **The ISO/IEC 27000 series of standards**, also known as the Information Security Management System (ISMS) family of standards, provides a systematic approach to managing sensitive company information and ensuring its confidentiality, integrity, and availability. The ISMS is built on a process of continuous improvement, which aligns well with the OODA Loop's emphasis on iterative learning and adaptation. By integrating the OODA Loop with the ISO/IEC 27000 series of standards, organizations can ensure that their cybersecurity risk management processes are robust, scalable, and in line with international best practices.

The OODA Loop is a useful tool for managing cybersecurity risks, but it is most effective when integrated with other frameworks and methodologies. By combining it with frameworks such as NIST, MITRE ATT&CK, and ISO/IEC 27000, organizations can create a comprehensive approach to cybersecurity risk management that better anticipates and responds to evolving threats.

The Role of Threat Intelligence in the OODA Loop: Enhancing the Observe and Orient Phases

In this discussion, we will delve into the role of threat intelligence in the OODA Loop, focusing on how it can provide timely, accurate, and actionable information about emerging threats and vulnerabilities, ultimately leading to a more effective and resilient cybersecurity posture.

The OODA Loop has proven to be a valuable tool in managing cybersecurity risks. One critical element in effectively implementing the OODA Loop in cybersecurity is the integration of threat intelligence, which can significantly enhance the Observe and Orient Phases.

Threat intelligence refers to the collection, analysis, and dissemination of information about potential or existing cyber threats and vulnerabilities. It encompasses a wide range of data sources, such as open-source intelligence (OSINT), social media intelligence (SOCMINT), human intelligence (HUMINT), technical intelligence (TECHINT), and intelligence from proprietary sources. By gathering and analyzing this information, organizations can gain insights into the tactics, techniques, and procedures (TTPs) employed by threat actors, as well as the vulnerabilities that may be exploited to launch cyberattacks.

- **Observe:** In the context of the OODA Loop, threat intelligence plays a pivotal role in the Observe Phase. This Phase involves collecting information about the cybersecurity environment, including details about potential threats, vulnerabilities, and the overall risk landscape. Threat intelligence can provide organizations with a wealth of information, enabling them to detect and identify cyber threats more effectively. By integrating threat intelligence into the Observe Phase, organizations can obtain a comprehensive view of the threat landscape, identify emerging trends, and better understand the context in which cyber threats occur.

The effective use of threat intelligence in the Observe Phase can help organizations identify potential threats and vulnerabilities earlier in the process, enabling them to take proactive measures to mitigate risks. For instance, an organization that is aware of a newly discovered vulnerability in a widely used software product can take steps to patch the vulnerability before threat actors can exploit it. Similarly, by being aware of the TTPs used by a specific threat actor or group, organizations can implement appropriate security controls and defenses to protect against those specific threats.

- **Orient:** Threat intelligence also plays a crucial role in the Orient Phase of the OODA Loop. During this Phase, organizations analyze the information gathered in the Observe Phase, considering their own unique context, resources, and objectives. This analysis helps organizations make sense of the information, understand the relevance and significance of the threats and vulnerabilities, and prioritize cybersecurity efforts accordingly.

Integrating threat intelligence into the Orient Phase can help organizations develop a more nuanced understanding of the risks they face, allowing them to make more informed decisions about their cybersecurity strategy. For example, by analyzing threat intelligence related to a specific industry or sector, organizations can gain insights into the most prevalent threats targeting their industry and prioritize their defenses accordingly. Similarly, by understanding the motivations and objectives of threat actors, organizations can better anticipate potential targets and implement appropriate security measures to protect their most critical assets.

The value of threat intelligence in the OODA Loop extends beyond the Observe and Orient Phases. By providing timely, accurate, and actionable information about emerging threats and vulnerabilities, threat intelligence can also inform the Decide and Act Phases.

- **Decide:** In the Decide Phase, organizations make decisions about the most appropriate actions to take based on their analysis and understanding of the threat landscape. Threat intelligence can help organizations make more informed decisions by providing them with up-to-date information about the risks they face, as well as the potential effectiveness of various security measures and controls.

- **Act:** In the Act Phase, organizations implement the decisions made in the Decide Phase, adjusting their cybersecurity strategy, defenses, and tactics as needed. Threat intelligence can play a critical role in this Phase as well, enabling organizations to respond more effectively and efficiently to emerging threats and vulnerabilities. By having access to relevant, actionable threat intelligence, organizations can quickly adapt their security measures and tactics in response to new or evolving threats, minimizing the potential impact of cyberattacks and improving their overall cybersecurity posture.

In addition to enhancing the individual Phases of the OODA Loop, threat intelligence can also facilitate more effective communication and collaboration among different stakeholders within an organization. By providing a common understanding of the threat landscape, threat intelligence can help break down silos and encourage cross-functional collaboration, ensuring that all relevant parties are working together to address cybersecurity risks. This collaboration can lead to more effective decision-making and action, as well as a more resilient and adaptive cybersecurity posture.

It is important to note, however, that simply collecting threat intelligence is not sufficient for enhancing the OODA Loop in cybersecurity. Organizations must also invest in the appropriate tools, processes, and personnel to effectively analyze, prioritize, and act on the information provided by threat intelligence. This includes implementing threat intelligence platforms and analytics tools, as well as developing a skilled workforce capable of interpreting and acting on the intelligence.

Moreover, organizations should strive to cultivate a culture of continuous learning and improvement, recognizing that the threat landscape is constantly evolving and that their cybersecurity strategies must adapt accordingly. By regularly reviewing and updating their OODA Loop processes considering new threat intelligence and changing circumstances, organizations can ensure that they maintain a proactive and adaptive approach to cybersecurity risk management.

The integration of threat intelligence into the OODA Loop can significantly enhance an organization's ability to detect, understand, and respond to cyber threats and vulnerabilities. By providing timely, accurate, and actionable information about emerging risks, threat intelligence can inform each Phase of the OODA Loop, ultimately leading to a more effective and resilient cybersecurity posture. To fully realize the benefits of threat intelligence in the OODA Loop, organizations must invest in the necessary tools, processes, and personnel, as well as cultivate a culture of continuous learning and improvement. With these elements in place, organizations can successfully navigate the complex and ever-changing world of cybersecurity, leveraging the power of the OODA Loop and threat intelligence to stay one step ahead of potential adversaries.

Automation and the OODA Loop

In this section, we will delve into the ways in which automation and AI can accelerate the OODA Loop process and enhance the overall effectiveness of cybersecurity efforts.

Automation and artificial intelligence (AI) have become increasingly prevalent in various aspects of our lives, including in the realm of cybersecurity. These advanced technologies have the potential to significantly impact the OODA Loop process, particularly in the Observe, Orient, and Act Phases, where tasks like threat detection, analysis, and remediation can benefit from automation.

- **Observe:** First, let us consider the role of automation and AI in the Observe Phase of the OODA Loop. In the context of cybersecurity, the Observe Phase involves the collection and monitoring of data related to an organization's networks, systems, and applications, as well as external sources of information, such as threat intelligence feeds and security news. The primary goal of this Phase is to detect potential cyber threats and vulnerabilities as quickly and accurately as possible, so that appropriate action can be taken.

The sheer volume of data generated in today's digital world can be overwhelming, making it difficult for human analysts to process and analyze relevant information in a timely manner. Automation and AI can be utilized to manage this burden. By leveraging advanced algorithms, machine learning, and other AI techniques, organizations can automate the process of collecting, filtering, and analyzing data in the Observe Phase. This can significantly speed up the detection of potential threats, as well as reduce the likelihood of false positives and negatives, ensuring that analysts can focus their attention on the most critical issues.

- **Orient:** In addition to improving the speed and accuracy of threat detection, automation and AI can also enhance the effectiveness of the Orient Phase in the OODA Loop. The Orient Phase involves analyzing the data collected in the Observe Phase, as well as other relevant information, to develop a comprehensive understanding of the current threat landscape and the organization's cybersecurity posture. This understanding is crucial for informing the Decision and Act Phases of the OODA Loop, as it helps organizations to prioritize their efforts and allocate resources in the most effective manner.

- The complex and constantly evolving nature of the threat landscape can make it challenging for human analysts to keep up with the latest trends and developments, as well as to identify patterns and correlations that may indicate a potential threat or vulnerability. Automation and AI can help to address these challenges by rapidly identifying patterns and trends during the processing of large volumes of data, and providing actionable insights that can inform the Orient Phase of the OODA Loop. This can enable organizations to adapt more quickly to new and emerging threats and vulnerabilities, as well as to make more informed decisions about their cybersecurity strategies and tactics.

- **Act:** The Act Phase of the OODA Loop can also benefit from automation and AI, particularly when it comes to threat remediation and incident response. In the event of a security incident, organizations must act quickly and decisively to contain the threat, minimize the damage, and recover their systems and data. This can involve a wide range of tasks, such as patching vulnerabilities, updating security configurations, blocking malicious IP addresses, and deploying new security measures.

Automation and AI can streamline and accelerate these tasks, ensuring that organizations can respond more effectively and efficiently to security incidents. For example, automated incident response platforms can detect and respond to security events in real-time, without the need for human intervention. These platforms can automatically apply security patches, update configurations, and execute other remediation actions, reducing the time it takes to contain and mitigate a threat. In addition, AI-driven security tools can predict and prevent potential threats before they materialize, further enhancing the effectiveness of the Act Phase in the OODA Loop.

However, it is important to recognize that automation and AI are not without their challenges and limitations. For example, these technologies can greatly enhance the speed and accuracy of threat detection and response.

OODA Loop in a distributed environment

In today's increasingly interconnected world, organizations are rapidly adopting remote work, cloud computing, and distributed systems to enhance operational efficiency and business agility. These modern technologies offer numerous benefits, such as reduced costs, increased scalability, and greater flexibility. However, they also introduce new and complex cybersecurity challenges that traditional security models may struggle to address effectively. In this context, it is crucial for organizations to adapt their cybersecurity strategies and leverage the OODA Loop to manage risks in these distributed environments effectively. By understanding and adapting the OODA Loop to the unique challenges posed by remote work, cloud computing, and distributed systems, organizations can develop a more robust and resilient cybersecurity posture.

Remote Work

First, let us examine the implications of remote work on cybersecurity and how the OODA Loop can be adapted to address these challenges. The shift towards remote work has expanded the attack surface for cyber adversaries, as employees access sensitive data and systems from various locations, often using personal devices and unsecured networks. This presents a myriad of new risks, such as phishing attacks, malware infections, and data breaches, which traditional security measures may be ill-equipped to handle.

To effectively manage cybersecurity risks in a remote work environment, organizations must adjust their Observe and Orient

Phases in the OODA Loop. This involves expanding their monitoring and data collection efforts to encompass remote endpoints, networks, and cloud services used by employees. By leveraging advanced security tools, such as endpoint detection and response (EDR) solutions, organizations can gain greater visibility into their remote workforce's activities and detect potential threats more quickly and accurately.

In addition to enhancing visibility, organizations must also adapt their Orient Phase to better understand the unique risks associated with remote work. This may involve updating threat intelligence feeds, incorporating data from remote devices and networks, and developing new analytical models to identify patterns and trends indicative of cyber threats in a distributed environment. By incorporating this information into their cybersecurity risk assessments, organizations can better prioritize their efforts and allocate resources to address the most significant risks.

The Decide and Act Phases of the OODA Loop must also be adapted to address the challenges posed by remote work. Organizations must develop new policies, procedures, and controls to secure their remote workforce, such as implementing multi-factor authentication, encrypting sensitive data, and providing regular security training for employees. Furthermore, organizations must ensure that their incident response plans are updated to account for the unique challenges posed by remote work, such as coordinating efforts across dispersed teams and addressing legal and regulatory considerations that may arise when dealing with cross-border incidents.

Cloud Computing

Next, let us consider the implications of cloud computing on cybersecurity and how the OODA Loop can be adapted to address these challenges. The adoption of cloud computing has fundamentally altered the way organizations store, process, and transmit data, often blurring the lines between traditional network boundaries and exposing new attack vectors for cyber adversaries. To effectively manage cybersecurity risks in a cloud computing environment, organizations must adapt their OODA Loop to account for the unique characteristics and challenges of this technology.

In the Observe Phase, organizations must expand their monitoring and data collection efforts to include cloud-based infrastructure, platforms, and applications. This may involve leveraging cloud-native security tools, such as Cloud Access Security Brokers (CASBs) and Cloud Security Posture Management (CSPM) solutions, which can provide real-time visibility into cloud-based assets and detect potential threats and vulnerabilities. Additionally, organizations must collaborate closely with their cloud service providers to ensure that appropriate security measures are in place and that relevant threats are met.

Measuring the effectiveness of the OODA Loop

"The definition of insanity is doing the same thing over and over again, but expecting different results." - Albert Einstein.

Metrics and KPI

In this section, we will investigate various methods for measuring the effectiveness of the OODA Loop in a cybersecurity program, including key performance indicators (KPIs) and metrics that can help organizations track and improve their OODA Loop implementation.

As organizations increasingly adopt the OODA Loop as part of their cybersecurity programs, it becomes essential to evaluate and measure the effectiveness of this decision-making framework in addressing and mitigating cyber threats and vulnerabilities. Understanding how well the OODA Loop is performing can enable organizations to make data-driven decisions, optimize their cybersecurity efforts, and ultimately enhance their overall security posture.

It is essential to recognize that the OODA Loop is not a static process, but rather an iterative and dynamic decision-making cycle. As such, measuring its effectiveness requires a comprehensive and continuous approach, considering various aspects of the organization's cybersecurity program, such as threat detection, response, and remediation. By collecting, analyzing, and reporting on a range of KPIs and metrics, organizations can develop a holistic understanding of their OODA Loop performance and identify areas for improvement. KPIs can focus on individual Phases, as well as the overall OODA Loop process.

- **Observe:** One approach to measuring the effectiveness of the OODA Loop is to examine the Observe Phase, focusing on the organization's ability to detect and identify potential cyber threats and vulnerabilities. Key metrics and KPIs in this Phase may include:

1. *Detection time:* The time it takes for the organization to detect a potential cyber threat or vulnerability from the moment it emerges. This metric can help organizations assess their monitoring capabilities and the effectiveness of their data collection efforts.

2. *False positive and false negative rates:* These rates refer to the number of incorrect threat detections (false positives) and missed threats (false negatives). High rates of either can indicate issues with the organization's threat detection capabilities and may necessitate further investigation or improvements.

3. *Coverage:* The percentage of the organization's assets, systems, and networks that are monitored and protected by its cybersecurity program. This metric can help organizations assess the comprehensiveness of their Observe Phase and identify potential gaps in their security coverage.

   - **Orient:** Next, organizations can evaluate the effectiveness of the Orient Phase of the OODA Loop, which focuses on analyzing collected data and developing a comprehensive understanding of the current threat landscape and the organization's cybersecurity posture. Metrics and KPIs in this Phase may include:

4. *Threat intelligence quality:* The accuracy, timeliness, and relevance of the threat intelligence used by the organization to inform its Orient Phase. High-quality threat intelligence can significantly enhance an organization's understanding of the threat landscape and improve decision-making in the OODA Loop.

5. *Time to analyze:* The time it takes for the organization to analyze collected data, identify patterns and trends, and develop actionable insights. Faster analysis times can indicate a more efficient Orient Phase and enable organizations to respond more quickly to emerging threats.

6. *Analytical false positives and false negatives:* These metrics measure the accuracy of the organization's analytical

processes in identifying true threats and vulnerabilities, as well as distinguishing them from false alarms. High rates of analytical false positives or false negatives can indicate issues with the organization's analytical capabilities and may necessitate further investigation or improvements.

- **Decide:** The Decide Phase of the OODA Loop involves determining the most appropriate course of action to address identified threats and vulnerabilities. To evaluate the effectiveness of this Phase, organizations can track metrics and KPIs, such as:

7. *Decision accuracy:* The proportion of correct decisions made by the organization in response to identified threats and vulnerabilities. High decision accuracy can indicate a well-functioning Decide Phase and help organizations allocate resources more effectively.

8. *Time to decide*: The time it takes for the organization to decide after analyzing the data and insights gathered during the Orient Phase. Faster decision-making times can enable organizations to respond more quickly to emerging threats, potentially minimizing their impact and reducing the likelihood of a successful attack.

9. *Decision agility*: The organization's ability to adapt and change its decisions in response to latest information or evolving threats. A high degree of decision agility can indicate a more effective Decide Phase and demonstrate the organization's ability to respond dynamically to changes in the threat landscape.

- **Act:** Finally, organizations can assess the effectiveness of the Act Phase of the OODA Loop, which focuses on executing the chosen course of action to address identified threats and vulnerabilities. Metrics and KPIs in this Phase may include:

10. *Time to remediate*: The time it takes for the organization to execute its chosen course of action and remediate identified threats and vulnerabilities. Faster remediation

times can help minimize the potential damage caused by a successful cyberattack and demonstrate the effectiveness of the organization's Act Phase.

11. *Remediation success rate*: The proportion of threats and vulnerabilities that are successfully remediated by the organization. A high remediation success rate can indicate a more effective Act Phase and help organizations maintain a strong security posture.

- **Post-incident analysis:** The organization's ability to learn from security incidents and improve its OODA Loop processes. By conducting thorough post-incident analyses, organizations can identify areas for improvement and implement changes to enhance the effectiveness of their OODA Loop implementation.

12. *Performance scorecards*: In addition to tracking individual metrics and KPIs, organizations can also benefit from developing an overall performance scorecard for their OODA Loop implementation. This scorecard can provide a high-level view of the organization's cybersecurity program's effectiveness, enabling decision-makers to assess progress, identify trends, and allocate resources more effectively.

It is important to note that the process of measuring the effectiveness of the OODA Loop in a cybersecurity program is not a one-time effort but should be conducted on an ongoing basis. By regularly monitoring and evaluating their OODA Loop performance, organizations can stay informed about the constantly evolving threat landscape, adapt their strategies and tactics accordingly, and ultimately improve their overall cybersecurity posture.

Measuring the effectiveness of the OODA Loop in a cybersecurity program is a critical component of maintaining a robust and resilient security posture. By tracking a range of key performance indicators and metrics across the Observe, Orient, Decide, and Act Phases, organizations can gain valuable insights into their OODA Loop implementation and identify areas for improvement. By continuously monitoring and evaluating their performance, organizations can ensure that their cybersecurity efforts are well-aligned with the dynamic and ever-changing threat landscape, enabling them to better protect their assets, systems, and networks from cyber threats and vulnerabilities.

Human factors in the OODA Loop

In this section, we will explore the influence of human factors such as cognitive biases, decision-making styles, and situational awareness on the OODA Loop and discuss their implications for cybersecurity professionals and organizations.

The OODA Loop is a critical decision-making process in cybersecurity, where organizations must constantly observe their environment, orient themselves to the situation, decide on a course of action, and act to protect their digital assets. While the process is often thought of as primarily technical, human factors also play a significant role in the successful application of the OODA Loop in cybersecurity.

Cognitive biases are systematic errors in thinking and decision-making that can influence how individuals perceive and process information. They can significantly impact the OODA Loop, especially in the Observe and Orient Phases, where individuals must collect and interpret data to make informed decisions. Examples of common cognitive biases that can affect the OODA Loop in cybersecurity include:

1. **Confirmation bias:** The tendency to seek out or interpret information in a way that confirms one's preexisting beliefs. In the context of cybersecurity, confirmation bias can lead to a false sense of security, as individuals may be more likely to focus on data that supports their current assumptions rather than seeking out potentially contradictory information.

2. **Availability bias:** The inclination to rely on readily available information rather than seeking out comprehensive data. In cybersecurity, this bias can lead to an overemphasis on recent or high-profile threats, potentially causing organizations to overlook less visible but equally significant risks.

3. **Anchoring bias:** The tendency to rely too heavily on an initial piece of information when making decisions. In the

cybersecurity context, this can result in security professionals being influenced by early assessments or estimates, potentially leading to inadequate responses to evolving threats.

4. **Groupthink:** The preference for consensus and conformity within a group, which can result in a reluctance to challenge prevailing opinions or consider alternative viewpoints. In cybersecurity, groupthink can hinder the effective implementation of the OODA Loop, as it may prevent security teams from critically evaluating their assumptions and exploring new strategies.

To mitigate the influence of cognitive biases on the OODA Loop, cybersecurity professionals should be aware of their potential impact and adopt strategies to counteract them. For example, organizations can encourage diversity of thought by fostering an open and inclusive culture where dissenting opinions are valued and considered. Additionally, security teams can use structured decision-making processes, such as red teaming or scenario planning, to challenge their assumptions and explore alternative perspectives.

Decision-making styles also play a significant role in the OODA Loop, as they can influence how individuals approach the Orient and Decide Phases. Common decision-making styles that can impact the OODA Loop in cybersecurity include:

1. **Analytical decision-making:** This style is characterized by a systematic and methodical approach to problem-solving, with individuals carefully considering all available information before deciding. While this style can lead to thorough and well-reasoned decisions in cybersecurity, it may also result in slower response times, which can be detrimental in rapidly evolving threat environments.

2. **Intuitive decision-making:** This style relies on gut instincts and intuition, with individuals making decisions quickly based on their feelings or past experiences. In the context of cybersecurity, intuitive decision-making can enable security professionals to respond rapidly to emerging threats;

however, it may also lead to impulsive or ill-considered actions that overlook crucial information.

3. **Collaborative decision-making:** This style emphasizes the importance of group input and consensus in decision-making. In cybersecurity, collaborative decision-making can foster a shared understanding of the threat landscape and promote more effective responses; however, it may also be subject to groupthink or slow decision-making processes.

4. **Decisive decision-making:** This style is characterized by a willingness to make quick decisions and commit to a course of action, even in the face of uncertainty or incomplete information. While decisive decision-making can be valuable in cybersecurity, allowing organizations to respond promptly to threats, it may also result in premature or overly aggressive actions that fail to account for potential consequences.

To optimize the OODA Loop in cybersecurity, organizations should understand the strengths and weaknesses of different decision-making styles and foster a balanced approach that combines the best aspects of each style. For instance, security teams can blend analytical and intuitive decision-making to ensure a comprehensive assessment of the situation while maintaining the ability to react quickly to emerging threats. Similarly, organizations can combine collaborative and decisive decision-making to encourage group input while ensuring timely and resolute actions.

Situational awareness is another critical human factor in the OODA Loop, as it refers to an individual's ability to perceive, comprehend, and predict the elements of their environment. In cybersecurity, situational awareness involves understanding the organization's digital assets, threat landscape, and potential vulnerabilities. It also includes the ability to recognize patterns and trends in the data and anticipate how the situation may evolve.

Developing strong situational awareness is crucial for the effective implementation of the OODA Loop in cybersecurity, as it enables security professionals to make informed decisions and act proactively in response to threats. Some strategies for enhancing situational awareness in cybersecurity include:

1. **Continuous monitoring and data collection:** Organizations should invest in advanced security tools and technologies that provide real-time visibility into their networks, systems, and applications. This can help security teams gather the necessary data to support the Observe and Orient Phases of the OODA Loop.

2. **Threat intelligence:** Cybersecurity professionals should stay up to date on the latest threat intelligence and incorporate it into their decision-making processes. This can help improve situational awareness by providing timely, accurate, and actionable information about emerging threats and vulnerabilities.

3. **Training and education:** Security teams should receive regular training and education on cybersecurity best practices, emerging threats, and the latest tools and technologies. This can help build the knowledge and skills necessary for maintaining situational awareness in a rapidly evolving threat landscape.

4. **Collaboration and information sharing:** Cybersecurity professionals should actively engage with their peers, both within their organization and across the broader cybersecurity community, to share insights and learn from one another's experiences. This can help enhance situational awareness by fostering a collective understanding of the threat landscape and promoting more effective responses.

The human factors of cognitive biases, decision-making styles, and situational awareness play a significant role in the successful application of the OODA Loop in cybersecurity. By understanding and addressing these factors, organizations can optimize their decision-making processes, enhance their cybersecurity posture, and better protect their digital assets against evolving threats.

Legal and Ethical Considerations in OODA Loop Implementation for Cybersecurity

This discussion provides a detailed examination of the legal and ethical dimensions of the OODA Loop in cybersecurity. We explore the a range of factors that organizations need to consider as they work to uphold the highest standards of professionalism and ethical conduct.

The implementation of the OODA Loop in cybersecurity comes with a myriad of legal and ethical considerations that organizations must address to effectively protect their systems, data, and users. These considerations stem from the increasing complexity of data protection laws and regulations, the need to respect individual privacy rights, and the importance of maintaining the integrity of intellectual property.

Compliance with Data Protection Laws and Regulations

As organizations around the world become more reliant on data to drive their operations, the importance of data protection has grown significantly. This heightened focus on data security has led to the development of numerous data protection laws and regulations aimed at safeguarding personal information and preserving user privacy. Among the most prominent of these are the European Union's General Data Protection Regulation (GDPR) and the California Consumer Privacy Act (CCPA) in the United States.

When implementing the OODA Loop in cybersecurity, organizations must ensure that they adhere to the provisions of these and other relevant data protection laws and regulations. This entails a comprehensive understanding of the legal requirements associated with the collection, processing, storage, and sharing of personal data. Compliance with these regulations requires that organizations develop and maintain robust data protection policies and procedures, including the designation of data protection officers, the establishment of processes for handling data subject requests, and the implementation of appropriate technical and organizational measures to secure personal information.

Failing to comply with data protection laws and regulations can result in severe penalties for organizations, including substantial fines and reputational damage. As such, it is imperative that cybersecurity professionals working within the framework of the OODA Loop prioritize regulatory compliance as a key element of their risk management strategies.

Respecting Privacy Rights and Intellectual Property

In addition to regulatory compliance, the implementation of the OODA Loop in cybersecurity also involves the consideration of individual privacy rights and the protection of intellectual property. Privacy rights, which are enshrined in various legal frameworks and ethical guidelines, are critical to ensuring that individuals maintain control over their personal information and can make informed decisions about how their data is used.

As cybersecurity professionals apply the OODA Loop to protect their organizations, they must ensure that their actions do not infringe upon the privacy rights of individuals. This means that security teams should strive to collect and process personal data in a manner that is transparent, lawful, and proportional to the identified security risks. Additionally, cybersecurity professionals must be mindful of the potential for unintentional privacy violations that may arise from the deployment of certain security tools or practices, such as the use of deep packet inspection or the collection of biometric data.

The protection of intellectual property is another important legal and ethical consideration in the context of the OODA Loop and cybersecurity. Intellectual property, which includes patents, trademarks, copyrights, and trade secrets, represents the valuable assets of individuals and organizations that drive innovation and economic growth. As such, cybersecurity professionals must be vigilant in their efforts to safeguard the intellectual property of their organizations and to respect the intellectual property rights of others.

This may involve implementing measures to prevent the unauthorized access, disclosure, or use of sensitive information, as well as the development of policies and procedures for managing third-party software and services. In this regard, cybersecurity professionals must be cognizant of the potential legal and ethical implications associated with the use of open-source software, the reverse engineering of proprietary technologies, and the sharing of threat intelligence data.

Balancing Security and Ethical Considerations

One of the primary challenges faced by cybersecurity professionals when implementing the OODA Loop is the need to strike a balance between security objectives and ethical considerations. This delicate balance requires a nuanced understanding of the potential risks and benefits associated with various security measures and the ability to make informed decisions that uphold the highest standards of ethical conduct.

To effectively navigate this balance, cybersecurity professionals should consider the following guiding principles:

1. **Proportionality:** Security measures should be proportional to the risks they are designed to mitigate. This means that organizations must carefully assess the potential impact of security tools and practices on individual privacy rights and intellectual property and ensure that they are not overly intrusive or invasive.

2. **Transparency:** Organizations should be transparent about their cybersecurity practices, including the types of data they collect and process, the purposes for which this data is used, and the measures they take to protect personal information and intellectual property. This transparency helps to build trust with users, customers, and regulators; and ensures that individuals can make informed decisions about their data.

3. **Accountability:** Cybersecurity professionals must be accountable for their actions and decisions, particularly when it comes to the implementation of the OODA Loop in

cybersecurity. This accountability extends to ensuring compliance with data protection laws and regulations, respecting individual privacy rights, and upholding the principles of ethical conduct.

4. **Collaboration:** Given the interconnected nature of the digital world, effective cybersecurity requires collaboration between organizations, governments, and industry partners. This collaboration should be grounded in a shared commitment to legal and ethical principles, and should involve the exchange of best practices, threat intelligence data, and other resources that can help to improve the overall security posture of all parties involved.

5. **Continuous improvement:** The rapidly evolving nature of the cybersecurity landscape means that organizations must constantly evaluate and update their security practices to stay ahead of emerging threats and vulnerabilities. This continuous improvement should be guided by a commitment to legal and ethical principles, as well as the lessons learned from past experiences and incidents.

By adhering to these guiding principles, cybersecurity professionals can ensure that their implementation of the OODA Loop in cybersecurity is consistent with the highest standards of legal and ethical conduct. This will not only help to protect their organizations from cyber threats, it will also serve to enhance public trust in the digital ecosystem and promote a more secure and resilient online environment for all.

The Role of Training and Education in the Effective Application of the OODA Loop in Cybersecurity

The rapidly evolving landscape of cybersecurity threats and challenges demands a workforce capable of responding effectively to these risks. The OODA Loop emphasizes the importance of rapid decision-making and adaptation to changing circumstances, which is particularly relevant in the dynamic field of cybersecurity.

In the context of the OODA Loop, training and education play a crucial role in ensuring that individuals involved in cybersecurity are well-equipped to apply this decision-making framework to protect their organizations. A well-trained and educated workforce is the backbone of any successful cybersecurity program and fostering a culture of continuous learning and improvement is essential to effectively apply the OODA Loop in the cybersecurity arena.

In the cybersecurity domain, the OODA Loop can be employed to respond to threats, manage incidents, and enhance the overall security posture of an organization. The ability to apply the OODA Loop effectively depends on the capabilities of the individuals involved, which is why training and education are essential components of any successful cybersecurity program. By investing in the development of their workforce, organizations can ensure that they are better prepared to face emerging threats and challenges.

Training and education in the context of the OODA Loop and cybersecurity can be divided into several areas, including foundational knowledge, skills development, advanced training, and ongoing learning. Each of these areas contributes to the overall effectiveness of the workforce in applying the OODA Loop and improving the cybersecurity posture of the organization.

1. **Foundational Knowledge:** A solid understanding of the OODA Loop principles and their relevance to cybersecurity is fundamental for individuals involved in the field. Foundational knowledge should cover the basics of the OODA Loop, its Phases, and its application in various

cybersecurity scenarios. This knowledge serves as the basis for further skills development and more advanced training. In addition to understanding the OODA Loop, foundational training should also encompass basic cybersecurity concepts, such as threat types, attack vectors, risk management, and security controls. This ensures that individuals have a broad understanding of the cybersecurity landscape, enabling them to apply the OODA Loop in a more informed and effective manner.

2. **Skills Development:** Once individuals have a solid grasp of the OODA Loop principles and basic cybersecurity concepts, the focus should shift to developing specific skills that are essential for applying the OODA Loop in practice. These skills can be divided into several categories, such as technical skills, analytical skills, and soft skills. Technical skills are essential for individuals involved in the hands-on aspects of cybersecurity, such as incident response, threat hunting, or security operations. Skills development in this area may include training in various cybersecurity tools, technologies, and platforms, as well as learning about different types of attacks, vulnerabilities, and defenses. Analytical skills are crucial for the Orient and Decide Phases of the OODA Loop, where individuals need to analyze the information gathered during the Observe Phase and make informed decisions. Training in this area may involve learning about various analytical techniques, data analysis tools, and methods for interpreting and visualizing information. Soft skills, such as communication, collaboration, and leadership, are also vital in the application of the OODA Loop in cybersecurity. As cybersecurity professionals work together to respond to threats and manage incidents, the ability to effectively communicate and collaborate with others is essential for a successful outcome. Training in this area may include workshops, role-playing exercises, and team-building activities designed to foster strong interpersonal skills and a collaborative mindset.

3. **Advanced Training:** As individuals progress in their cybersecurity careers and gain experience in applying to the OODA Loop, they may benefit from more advanced training focused on specific aspects of the framework or specialized areas of cybersecurity. This advanced training can help to deepen their understanding of the OODA Loop and its application in various contexts, as well as sharpen their skills in areas such as threat intelligence, digital forensics, or incident management. Advanced training can also include exploring topics such as the integration of the OODA Loop with other cybersecurity frameworks and methodologies, the role of automation and artificial intelligence in accelerating the OODA Loop process, or the adaptation of the OODA Loop for distributed environments and remote work scenarios.

4. **Ongoing Learning:** The cybersecurity landscape is constantly evolving, and new threats and challenges emerge on a regular basis. To remain effective in applying the OODA Loop in this dynamic environment, individuals must engage in ongoing learning and continuous improvement. This can include participating in industry conferences, webinars, workshops, or online courses, as well as staying up to date with the latest research, trends, and best practices in cybersecurity. Ongoing learning also involves maintaining an awareness of the legal and ethical considerations surrounding cybersecurity, as well as the importance of resilience, adaptability, and effective communication in the successful application of the OODA Loop. By fostering a culture of continuous learning, organizations can ensure that their cybersecurity workforce remains agile, adaptable, and prepared to face emerging challenges.

The importance of training and education as critical factors in the effective application of the OODA Loop in cybersecurity cannot be overemphasized. By investing in the development of their workforce, organizations can build a solid foundation for their cybersecurity program and enhance their ability to respond to threats, manage incidents, and improve their overall security posture. A well-trained and educated workforce is the backbone of any successful cybersecurity program and fostering a culture of continuous learning and improvement is essential to effectively apply the OODA Loop in the cybersecurity arena. By focusing on foundational knowledge, skills development, advanced training, and ongoing learning, organizations can empower their employees to apply the OODA Loop effectively and protect their organizations from the ever-evolving landscape of cybersecurity threats and challenges.

Resilience and Adaptability

Resilience and adaptability are critical components of any effective cybersecurity strategy. In the context of the OODA Loop, these characteristics are particularly essential, as they enable organizations to continuously observe, orient, decide, and act in response to the ever-changing threat landscape. By embracing resilience and adaptability, organizations can not only better protect themselves against current threats but also develop the flexibility needed to respond to future challenges.

One of the primary reasons why resilience and adaptability are so important in the realm of cybersecurity is the dynamic nature of the threat landscape. Cybersecurity threats are constantly evolving, with new vulnerabilities, attack vectors, and threat actors emerging on a regular basis. This rapid pace of change means that organizations cannot afford to be complacent or to rely on static security measures that may quickly become outdated or ineffective.

To build resilience and adaptability into their cybersecurity strategies, organizations must focus on several key areas:

1. **Continuous monitoring and assessment:** A robust cybersecurity program requires constant vigilance, with organizations monitoring their networks, systems, and data for signs of potential threats or compromises. This continuous monitoring should involve the use of both automated tools, such as intrusion detection systems, open-source intelligence gathering platforms, and security

information and event management (SIEM) platforms, as well as manual analysis and review by skilled security professionals. By maintaining a constant watch over their digital assets, organizations can more quickly identify and respond to potential threats.

2. **Threat intelligence:** An integral component of resilience and adaptability is staying informed about the latest threats and vulnerabilities that could potentially impact an organization's security posture. This involves the collection, analysis, and dissemination of threat intelligence, which can be derived from a variety of sources, including industry reports, security blogs, government advisories, and information sharing platforms. By staying abreast of the latest threats, organizations can better anticipate and prepare for potential attacks.

3. **Regular policy and process review:** Security policies and processes should be subject to regular review and updates to ensure that they remain relevant and effective in the face of new challenges. This may involve revising access control policies, updating incident response plans, or reevaluating the organization's risk management framework. By regularly reviewing and updating these elements, organizations can ensure that they are well-prepared to address emerging threats and vulnerabilities.

4. **Training and awareness:** To maintain resilience and adaptability, it is essential that all employees within an organization are well-trained and aware of the latest cybersecurity threats and best practices. This includes not only technical staff but also non-technical employees who may be targeted by social engineering attacks or other forms of deception. By fostering a culture of security awareness and providing ongoing training, organizations can help to ensure that their employees are well-equipped to recognize and respond to potential threats.

5. **Technology updates and upgrades:** As new threats emerge and the cybersecurity landscape evolves, it is crucial that organizations remain at the forefront of security technology

to better defend themselves against the constantly evolving threat landscape. This may involve upgrading existing security tools, such as firewalls and antivirus software, as well as investing in novel solutions that can provide enhanced protection against emerging threats. Testing and validation: A key aspect of resilience and adaptability is the ability to test and validate the effectiveness of an organization's security measures. This may involve conducting regular vulnerability assessments, penetration tests, and other forms of security testing to identify potential weaknesses and areas for improvement. By rigorously testing their security posture, organizations can gain valuable insights into their strengths and vulnerabilities, enabling them to make informed decisions about how to enhance their defenses.

6. **Incident response planning:** Despite the best efforts of security professionals, it is virtually impossible to prevent every potential cyberattack or security breach. As such, organizations must develop comprehensive incident response plans that outline the steps to be taken in the event of a security incident. These plans should include clear roles and responsibilities for all team members, as well as processes for identifying, containing, and remediating security incidents. By having a well-defined incident response plan in place, organizations can more effectively manage security events and minimize their potential impact.

8. **Learning from past incidents:** To build resilience and adaptability, organizations must learn from past security incidents and use these experiences to inform their future cybersecurity strategies. This involves conducting thorough post-mortem analyses of security events, identifying the root causes of the incidents, and developing strategies for preventing similar occurrences in the future. By learning from past mistakes and continually refining their cybersecurity strategies, organizations can better protect themselves against future threats.

9. **Collaborating with external partners:** Cybersecurity is a shared responsibility, and organizations can benefit from collaborating with external partners, such as industry peers, government agencies, and security vendors. This can involve participating in information sharing initiatives, leveraging shared threat intelligence, and collaborating on the development of new security technologies and best practices. By working together with other stakeholders, organizations can more effectively address the complex and evolving challenges posed by the cybersecurity threat landscape.

10. **Embracing a risk-based approach:** To be truly resilient and adaptable, organizations must adopt a risk-based approach to cybersecurity. This involves identifying and prioritizing the most significant risks facing the organization and allocating resources accordingly. By focusing on the most critical risks, organizations can ensure that their security efforts are targeted and effective, enabling them to adapt more effectively to the constantly changing threat landscape.

11. **Continuous improvement:** Resilience and adaptability are not one-time efforts; rather these are ongoing processes that require continuous improvement. Organizations should constantly strive to enhance their cybersecurity posture, leveraging innovative technologies, best practices, and insights gained from past experiences to strengthen their defenses. This commitment to continuous improvement will help organizations stay one step ahead of emerging threats and maintain a strong security posture in the face of an ever-changing landscape.

Resilience and adaptability are critical components of a successful cybersecurity strategy, particularly in the context of the OODA Loop. By focusing on continuous monitoring and assessment, threat intelligence, regular policy and process review, training and awareness, technology updates and upgrades, testing and validation, incident response planning, learning from past incidents, collaborating with external partners, embracing a risk-based approach, and fostering continuous improvement, organizations can more effectively manage their cybersecurity risks and navigate the complex and dynamic threat landscape.

Communication and collaboration

This section summarizes key elements analyzed throughout this book, which together can be viewed to demonstrate how effective communication and collaboration play a crucial role in the successful implementation of the OODA Loop in cybersecurity. To fully capitalize on the potential of the OODA Loop, it is essential for organizations to foster a culture of open communication and collaboration, not only within the security team but also with other stakeholders, such as IT teams, business units, management, and external partners. By establishing clear channels for sharing information and insights, organizations can enhance their collective understanding of the threat landscape and improve their overall cybersecurity posture.

1. **Internal communication and collaboration:** The OODA Loop relies on the rapid flow of information through the Observe, Orient, Decide, and Act Phases, which requires seamless communication and collaboration among various members of the security team and other stakeholders within the organization. To facilitate this flow of information, organizations should establish well-defined communication channels and processes that enable team members to share relevant data, insights, and updates quickly and effectively. This can include the use of collaboration tools, such as chat applications, ticketing systems, and shared dashboards, as well as regular meetings and briefings to keep all stakeholders informed and engaged.

2. **Cross-functional collaboration:** Cybersecurity is not solely the responsibility of the security team but rather a shared responsibility that spans across the entire organization. To fully leverage the OODA Loop, it is essential for security teams to work closely with other departments, such as IT, human resources, legal, and management. This cross-

functional collaboration can help organizations ensure that their cybersecurity strategies and initiatives are aligned with their overall business objectives and that all stakeholders are working together to protect the organization from potential threats.

3. **Security awareness and training:** Communication and collaboration within the OODA Loop also involve raising awareness and providing training to employees at all levels of the organization. As the human factor is often cited as the weakest link in cybersecurity, it is essential for organizations to invest in security awareness and training programs that equip employees with the knowledge and skills needed to identify and respond to potential threats. By fostering a security-conscious culture, organizations can reduce the likelihood of successful attacks and improve their overall cybersecurity posture.

4. **Sharing threat intelligence:** The Observe and Orient Phases of the OODA Loop rely heavily on the availability and analysis of accurate and timely threat intelligence. By sharing threat intelligence both internally and externally, organizations can gain a more comprehensive understanding of the threat landscape and enhance their ability to respond effectively to potential threats. This can involve participation in industry-specific information sharing and analysis centers (ISACs), as well as collaboration with government agencies, security vendors, and other external partners.

5. **Collaborating on incident response:** In the event of a security incident, effective communication and collaboration become even more critical. Organizations should have well-defined incident response plans in place, outlining the roles and responsibilities of various stakeholders and establishing clear lines of communication. By working together in a coordinated manner, organizations can more effectively contain and remediate security incidents, minimizing their potential impact and reducing the time required to recover.

6. **Engaging with external partners:** In addition to internal communication and collaboration, organizations should also engage with external partners to enhance their cybersecurity capabilities. This can include working with security vendors and consultants to access specialized expertise, tools, and services, as well as collaborating with industry peers, government agencies, and other stakeholders to share best practices, threat intelligence, and other valuable resources.

7. **Participating in industry forums and events:** Another way to foster communication and collaboration within the OODA Loop is to participate in industry forums and events, such as conferences, workshops, and webinars. These events can provide valuable opportunities for organizations to learn from the experiences of others, share their own insights and lessons learned, and network with potential partners and collaborators.

8. **Leveraging professional networks and communities:** Cybersecurity professionals can also benefit from engaging with their professional networks and communities, such as industry associations, online forums, and social media groups. By actively participating in these networks and communities, professionals can stay informed about the latest trends, technologies, and threats, as well as share their own experiences and insights with peers. This can help foster a collaborative mindset that enhances the application of the OODA Loop in cybersecurity.

9. **Promoting a culture of transparency and trust:** For communication and collaboration to be effective within the OODA Loop, it is vital to establish a culture of transparency and trust within the organization. This means fostering an environment where employees feel comfortable sharing information about potential threats and vulnerabilities without fear of reprisal or blame. By promoting open and honest dialogue, organizations can encourage the sharing of information and insights that can help enhance the effectiveness of the OODA Loop in cybersecurity.

10. **Encouraging innovation and continuous improvement:** Communication and collaboration are also essential for driving innovation and continuous improvement within the OODA Loop. By encouraging team members to share their ideas, experiences, and insights, organizations can identify opportunities for improvement and develop new strategies, tactics, and tools to strengthen their cybersecurity posture. This can involve implementing feedback loops, organizing brainstorming sessions, and creating mechanisms for the ongoing evaluation and refinement of the OODA Loop process.

11. **Collaborative learning and knowledge sharing:** One of the key benefits of effective communication and collaboration within the OODA Loop is the opportunity for collaborative learning and knowledge sharing. By working together to analyze and understand the threat landscape, organizations can pool their collective expertise and resources to develop more robust and resilient cybersecurity defenses. This can involve creating and maintaining shared knowledge bases, organizing training sessions and workshops, and leveraging the power of collective intelligence to stay ahead of emerging threats.

12. **Aligning cybersecurity and business objectives:** Effective communication and collaboration within the OODA Loop can also help organizations align their cybersecurity strategies and initiatives with their overall business objectives. By engaging with stakeholders and ensuring that their concerns and priorities are considered, organizations can develop a more holistic and integrated approach to cybersecurity risk management that supports their strategic goals.

Communication and collaboration are foundational factors in the successful application of the OODA Loop in cybersecurity. By fostering a culture of openness, transparency, and cooperation, organizations can enhance their collective understanding of the threat landscape, improve their overall cybersecurity posture, and better protect themselves from potential threats. This requires not only the establishment of clear channels for sharing information and insights but also a commitment to ongoing learning, innovation, and continuous improvement that ensures the OODA Loop remains effective in the face of ever-evolving cybersecurity challenges.

Post-incident analysis and learning

Post-incident analysis and learning are essential to an effective cybersecurity strategy, as they enable organizations to continually refine their processes and adapt their defenses in response to evolving threats. By conducting thorough reviews of security incidents and their handling, security teams can identify valuable lessons and insights that can help them improve their application of the OODA Loop and enhance their overall cybersecurity posture. In this context, post-incident analysis and learning can be seen as a critical feedback mechanism that supports the ongoing optimization of the OODA Loop, ensuring that organizations are better prepared to respond to future threats and minimize the potential impact of security breaches.

1. **The importance of post-incident analysis:** The primary goal of post-incident analysis is to understand the root causes of security incidents and their consequences, as well as the effectiveness of the organization's response. This involves examining various aspects of the incident, such as the tactics, techniques, and procedures (TTPs) used by the attackers, the vulnerabilities that were exploited, the organization's detection and response capabilities, and the overall impact of the incident on the business. By gaining a deep understanding of these factors, organizations can identify areas for improvement and develop actionable recommendations that can help them strengthen their defenses and reduce the likelihood of future incidents.

2. **Key elements of post-incident analysis:** A comprehensive post-incident analysis should include several key elements, such as the following:

- *Timeline of events*: A detailed timeline of the incident, from the initial point of compromise to the resolution, can help security teams understand the sequence of events and identify any gaps or delays in their response.
- *Attack vectors and TTPs*: An analysis of the attack vectors and TTPs used by the threat actors can help organizations understand the nature of the threat and the methods used to breach their defenses.
- *Vulnerabilities and weaknesses*: Identifying the vulnerabilities and weaknesses that were exploited during the incident can help organizations prioritize their remediation efforts and ensure that they are addressing the most critical risks.
- *Detection and response capabilities*: Assessing the organization's detection and response capabilities can help identify any shortcomings in their ability to quickly identify and respond to security incidents.
- *Impact assessment*: Evaluating the overall impact of the incident on the organization, including financial, operational, and reputational damage, can help security teams understand the potential consequences of security breaches and prioritize their risk mitigation efforts.

3. **Structured post-incident analysis methodologies:** To ensure that post-incident analysis is thorough and effective, organizations should adopt structured methodologies that provide a systematic approach to the review process. These methodologies can help security teams ensure that they are considering all relevant aspects of the incident and its handling, as well as providing a consistent framework for the evaluation and comparison of different incidents. Examples of structured post-incident analysis methodologies include:

- *After Action Reviews (AARs)*: AARs are a structured process used by the military and other organizations to review and learn from training exercises, operations, and incidents. The AAR process typically involves a facilitated discussion among the participants, focusing on the objectives, actions, and outcomes of the incident and identifying areas for improvement.

- *Root Cause Analysis (RCA)*: RCA is a problem-solving method that focuses on identifying the underlying causes of incidents, rather than just addressing the immediate symptoms. RCA involves a systematic process of data collection, analysis, and synthesis, with the aim of identifying the root causes of incidents and developing recommendations for their prevention.

- *Incident Response Life Cycle (IRLC)*: The IRLC is a model that describes the various Phases of an incident response process, from preparation and detection to containment, eradication, recovery, and post-incident analysis. By using the IRLC as a framework for post-incident analysis, organizations can ensure that they are considering all relevant aspects of the incident and its handling.

4. **Learning from post-incident analysis:** The goal of post-incident analysis is to enable organizations to learn from their experiences and improve their security practices. To achieve this, it is important that the lessons and insights gained from the analysis are effectively communicated and integrated into the organization's cybersecurity strategy. This can involve several steps, such as:

- *Sharing findings and recommendations*: Security teams should ensure that the findings and recommendations from their post-incident analysis are shared with relevant stakeholders, such as senior management, IT teams, and other business units. This can help build awareness of the risks and challenges faced by the organization and foster a culture of continuous improvement.

- *Updating policies, processes, and technologies*: Based on the insights gained from post-incident analysis, organizations should review and update their security policies, processes, and technologies to ensure that they remain effective in the face of evolving threats. This may involve making changes to their network architecture, implementing new security controls, or adopting new tools and technologies.

- *Training and awareness programs*: Organizations should use the lessons learned from security incidents to enhance their training and awareness programs, ensuring that employees are aware of the latest threats and best practices for protecting against them. This can help reduce the likelihood of future incidents by improving the overall security culture within the organization.

- *Continuous monitoring and improvement*: Security teams should establish processes for the ongoing monitoring and evaluation of their security practices, using metrics and indicators to track their performance and identify areas for improvement. By continually refining their OODA Loop implementation and learning from their experiences, organizations can become more agile and resilient in the face of cybersecurity threats.

5. **The role of post-incident analysis in the OODA Loop:** Post-incident analysis can be seen as an essential feedback mechanism within the OODA Loop, helping organizations refine their Observe, Orient, Decide, and Act processes and improve their overall cybersecurity posture. By conducting regular post-incident analysis and integrating the lessons learned into their OODA Loop, organizations can ensure that they are continually adapting and evolving their defenses in response to the changing threat landscape. This can help them maintain a proactive and resilient approach to cybersecurity, reducing the likelihood of future incidents and minimizing the potential impact of security breaches.

Post-incident analysis and learning are critical components of the OODA Loop that enable organizations to continually improve their cybersecurity practices and adapt their defenses in response to evolving threats. By conducting thorough reviews of security incidents and their handling, security teams can identify valuable lessons and insights that can help them refine their OODA Loop implementation and enhance their overall cybersecurity posture. By prioritizing post-incident analysis and learning, organizations can ensure that they are better prepared to respond to future threats and minimize the potential impact of security breaches.

Leadership and organizational culture in the OODA Loop implementation

This section summarizes those elements in the success of implementing the OODA Loop in cybersecurity that leaders can adopt, as this success largely depends on leadership and the organizational culture. Leaders play a crucial role in fostering a culture that values adaptability, continuous learning, and effective communication. By setting the tone for the organization, leaders can ensure that the OODA Loop is embraced across the organization and that teams are aligned towards a common goal. Some aspects to consider are:

- **Encouraging innovation and experimentation:** Leaders should promote a culture that encourages innovation and experimentation within the organization, allowing teams to test innovative ideas and learn from their successes and failures. This can help organizations stay agile and adapt quickly to the changing threat landscape.

- **Promoting collaboration and information sharing:** Effective collaboration and information sharing are essential to the OODA Loop implementation. Leaders should create an environment where teams are encouraged to share their insights, knowledge, and experiences, both within and outside the organization. This can help enhance collective understanding and improve the overall cybersecurity posture.

- **Reinforcing a learning mindset:** Leaders should emphasize the importance of continuous learning and professional development, helping employees stay up to date with the latest threats, best practices, and technological advancements. This can help build a more skilled and resilient workforce capable of responding effectively to cybersecurity challenges.

Balancing proactive and reactive cybersecurity strategies

Implementing the OODA Loop in cybersecurity requires organizations to strike the right balance between proactive and reactive strategies. While the OODA Loop is primarily focused on enabling organizations to respond quickly and effectively to threats, it is also essential to invest in proactive measures that can help prevent incidents from occurring in the first place. Some aspects to consider include:

- **Threat hunting and proactive defense:** Organizations should invest in threat hunting and proactive defense measures, such as continuous monitoring, vulnerability assessments, and penetration testing, to identify and address potential threats before they can cause damage. This can help organizations stay one step ahead of adversaries and reduce the likelihood of security incidents.
- **Cybersecurity risk management:** Implementing a robust cybersecurity risk management program can help organizations identify, assess, and prioritize their risks, enabling them to allocate resources more effectively and focus on the most critical threats. This can help organizations build a more resilient cybersecurity posture and ensure that their OODA Loop implementation is focused on addressing the most significant risks.
- **Employee training and awareness:** Investing in employee training and awareness programs can help organizations build a strong security culture and reduce the likelihood of incidents caused by human error. By equipping employees with the knowledge and skills they need to recognize and respond to threats, organizations can strengthen their defenses and improve their overall cybersecurity posture.

OODA and FAIR (Factor Analysis of Information Risk)

"FAIR gives organizations a way to prioritize their remediation efforts based on the risk reduction each one will bring. It helps you identify your top risks, and then helps you identify what you can do to reduce those risks in a cost-effective way." - Jack Jones

FAIR (Factor Analysis of Information Risk) is a quantitative risk assessment methodology that helps organizations identify and prioritize information security risks. It involves analyzing and quantifying factors such as asset value, threat frequency, threat capability, and control effectiveness. The OODA Loop, on the other hand, is a decision-making framework that emphasizes speed and agility in responding to changing situations.

When conducting a FAIR assessment, the OODA Loop can be used to help the organization respond quickly to any identified risks. During the Observe Phase, the organization can collect and analyze data on the factors that contribute to the risk, such as the likelihood of a threat occurring and the potential impact on assets. In the Orient Phase, the organization can use this information to develop a comprehensive understanding of the risk and identify possible courses of action. In the Decide Phase, the organization can select the most appropriate response strategy based on the data analyzed during the Observe and Orient Phases. Finally, in the Act Phase, the organization can implement the selected response strategy and monitor the results to determine whether additional action is necessary.

The OODA Loop can help organizations to be more efficient and effective in their risk assessment efforts. By breaking down the process into smaller Phases, the organization can make more informed decisions and respond quickly to any identified risks. This approach can help organizations to proactively manage risks and minimize potential impacts on their operations. However, it is important to note that the OODA Loop is not a replacement for the rigorous analysis required by the FAIR methodology. Rather, it is a complementary framework that can help organizations to make more effective decisions based on the data they have collected.

Case Study – 10.1 Utilizing OODA-FAIR

Sarah is a cybersecurity analyst working for a financial institution. One of her responsibilities is to assess the risk associated with their third-party vendors, which have access to the financial institution's sensitive data. As part of the assessment process, Sarah decides to use the Factor Analysis of Information Risk (FAIR) methodology, which allows her to quantify the risks associated with each vendor. In addition to FAIR, Sarah also decides to apply the OODA Loop to her assessment process to enhance the speed and effectiveness of her analysis.

1. **Observe:** Sarah begins her assessment by gathering all the necessary information about the third-party vendor. This includes the vendor's reputation, the type of data they have access to, their security controls, and any past security incidents they may have experienced. Sarah then analyzes this information to identify any potential risks associated with the vendor. Using the OODA Loop, Sarah can quickly identify the critical information she needs and streamline her data collection process.

2. **Orient:** Once Sarah has collected all the relevant information, she begins to analyze it in the context of her organization's risk appetite and business objectives. She considers the potential impact of a data breach on the financial institution and the likelihood of it occurring. Sarah also considers the vendor's security controls and their effectiveness in mitigating the identified risks. With the OODA Loop, Sarah can quickly assess the potential risks and determine the most appropriate course of action.

3. **Decide:** Based on her analysis, Sarah identifies several potential risks associated with the third-party vendor. She decides to recommend that additional security controls be implemented to mitigate these risks. Sarah also suggests that a contingency plan be developed in case a data breach was to occur. With the OODA Loop, Sarah can make informed decisions quickly and effectively.

4. **Act:** Finally, Sarah takes action to implement the security controls and contingency plan she

recommended. She works with the vendor to ensure that they are aware of the identified risks and collaborates with her team to implement the necessary controls. With the OODA Loop, Sarah can act quickly to reduce the risks associated with the vendor and ensure the safety of the financial institution's data.

In conclusion, the OODA Loop was an effective tool for Sarah to use in conjunction with the FAIR methodology to assess the risks associated with the third-party vendor. It allowed her to quickly gather and analyze critical information, make informed decisions, and take action to mitigate potential risks. By incorporating the OODA Loop into her assessment process, Sarah was able to enhance the speed and effectiveness of her analysis and ultimately reduce the risk of a data breach occurring.

CHAPTER 11: What About "The Art of War" and OODA?

"The supreme art of war is to subdue the enemy without fighting." - Sun Tzu

Early in my career I was captivated by Sun Tzu's seminal work, "The Art of War." The timeless wisdom and strategic thinking principles found in this ancient text significantly influenced my approach to problem-solving and decision-making. Sun Tzu's emphasis on understanding oneself and one's enemy, exploiting opportunities, and employing deception and indirect approaches resonated deeply with me. As I progressed in my career, these lessons from "The Art of War" formed the foundation of my strategic thinking, guiding me in the development and implementation of innovative solutions to complex challenges. My early exposure to this powerful treatise not only shaped my personal approach to strategy but also helped pave the way for my later interest in and application of the OODA Loop in cybersecurity.

Sun Tzu's "The Art of War" is a timeless work on military strategy that has influenced countless leaders as well as me, and thinkers across various fields, including cybersecurity. While the OODA Loop was developed by John Boyd, a military strategist and pilot, it is evident that Sun Tzu's teachings may have had an impact on Boyd's thinking and the development of the OODA Loop concept. In this analysis, we will explore several key principles from "The Art of War" and discuss how they may have influenced the OODA Loop framework.

1. **Speed and Adaptability:** Sun Tzu emphasized the importance of speed and adaptability in warfare. He believed that those who can adapt to changing circumstances and respond quickly to new information would have a significant advantage over their adversaries. This principle aligns well with the OODA Loop's focus on rapid decision-making and adaptation. The OODA Loop is designed to help organizations process information quickly,

make decisions, and take action in a fast-paced, ever-changing environment, such as cybersecurity.

2. **Knowing Your Enemy and Yourself:** Sun Tzu famously wrote, "If you know the enemy and know yourself, you need not fear the result of a hundred battles." This idea of understanding both your own strengths and weaknesses and those of your opponent is reflected in the OODA Loop's Orient Phase. During this Phase, organizations analyze, and process information gathered about their environment, including potential threats and vulnerabilities, as well as their own capabilities and limitations.

3. **Deception and Indirect Approaches:** Sun Tzu often stressed the value of deception and indirect approaches in warfare. He argued that an indirect approach could confuse and demoralize the enemy, leading to their defeat. In the context of the OODA Loop, this concept may be applied to the development of countermeasures and strategies that exploit the adversary's weaknesses or mislead them, ultimately protecting the organization's assets.

4. **Strategic Positioning:** Sun Tzu believed that strategic positioning is vital in warfare, as it enables an army to leverage its strengths and exploit the enemy's weaknesses. This concept can be applied to the OODA Loop by positioning the organization to respond effectively to cybersecurity threats, such as investing in the right technologies and developing robust incident response plans.

5. **Exploiting Opportunities:** Sun Tzu emphasized the importance of being opportunistic in warfare, taking advantage of favorable conditions and timing to strike the enemy. In the OODA Loop, organizations must be able to identify and exploit opportunities in the cybersecurity landscape, such as emerging vulnerabilities or weaknesses in an attacker's infrastructure.

6. **Unity of Effort and Command:** Sun Tzu stressed the importance of unity of effort and command, where all members of an army work together under a single, cohesive strategy. This principle is applicable to the OODA Loop in

the context of cybersecurity, where collaboration and communication among various stakeholders are critical for success.

Sun Tzu's "The Art of War" clearly had an influence on John Boyd's development of the OODA Loop concept. "He owned seven thoroughly annotated copies of the Art of War" (**https://fs.blog/ooda-loop/**) The principles of speed and adaptability, knowing your enemy and yourself, deception and indirect approaches, strategic positioning, exploiting opportunities, and unity of effort and command are all evident in the OODA Loop framework. By incorporating these timeless lessons from Sun Tzu, the OODA Loop can serve as an effective tool for organizations seeking to navigate the complex and ever-evolving world of cybersecurity.

CHAPTER 12: Deeper Dive - Speed and Adaptability in the Context of the OODA Loop and Cybersecurity

In today's fast-paced and rapidly evolving cybersecurity landscape, the importance of speed and adaptability cannot be overstated. These two key principles, which are deeply rooted in Sun Tzu's "The Art of War," are also fundamental components of John Boyd's OODA Loop. By understanding how speed and adaptability apply to the OODA Loop and cybersecurity, organizations can better navigate the complex threat environment and protect their assets.

Speed in the OODA Loop and Cybersecurity

In the context of the OODA Loop, speed refers to the ability of an organization to rapidly progress through the four Phases of the loop: Observe, Orient, Decide, and Act. The faster an organization can move through these Phases, the better equipped it is to respond to cybersecurity threats and incidents. Speed is crucial for several reasons:

- **Rapid threat detection:** The sooner an organization can identify a potential threat or security incident, the more likely it is to mitigate the impact or prevent the attack altogether. Speed in the Observe Phase allows organizations to quickly gather and process relevant information, enabling them to identify and address potential threats before they can cause significant damage.
- **Swift decision-making:** Once a threat has been identified, organizations must make timely decisions on how to respond. In the Decide Phase, speed is critical to ensure that appropriate actions are taken before an attacker can further exploit vulnerabilities or compromise valuable assets.
- **Prompt action:** Speed in the Act Phase is essential to quickly implement countermeasures and remediation efforts, reducing the overall impact of a cybersecurity incident.

Rapid action can limit the damage caused by an attack, safeguard sensitive information, and restore the organization's normal operations.

Adaptability in the OODA Loop and Cybersecurity

Adaptability is the capacity of an organization to adjust its strategies, tactics, and processes in response to changing circumstances and new information. In the context of the OODA Loop and cybersecurity, adaptability is vital for several reasons:

- **Evolving threat landscape:** Cyber threats are continually changing, with new attack vectors, vulnerabilities, and techniques emerging regularly. Adaptability in the Orient Phase enables organizations to stay up to date with the latest threat intelligence, ensuring that their security posture remains robust and effective.
- **Learning from experiences:** As organizations encounter various cybersecurity incidents, they can learn from these experiences to improve their OODA Loop execution. By adapting their strategies and processes based on lessons learned, organizations can enhance their ability to respond to future threats.
- **Resilience:** An adaptable organization is more resilient in the face of cybersecurity challenges. By being able to adjust to new threats and rapidly implement changes, organizations can better withstand attacks and recover from incidents.
- **Innovation:** Adaptability also encourages organizations to embrace innovation and explore new tools, technologies, and approaches to enhance their cybersecurity posture. By continually refining and updating their strategies, organizations can stay ahead of adversaries and maintain a strong defense.

Speed and adaptability are critical components of the OODA Loop and are essential for effective cybersecurity. By embracing these principles and integrating them into their cybersecurity strategies, organizations can better protect themselves from the ever-evolving threat landscape and ensure the security of their valuable assets.

Knowing Your Enemy and Yourself: The Importance of Strategic Insight in the OODA Loop and Cybersecurity

In the realm of cybersecurity, one of the most crucial factors for success is an in-depth understanding of both one's own organization and the adversaries that pose threats. This principle of knowing your enemy and yourself is derived from Sun Tzu's "The Art of War" and is a fundamental concept in John Boyd's OODA Loop. By cultivating strategic insight into the strengths and weaknesses of both internal and external elements, organizations can enhance their cybersecurity posture and better defend against attacks.

Knowing Your Enemy

Developing a comprehensive understanding of the enemy in the context of cybersecurity involves several key aspects:

1. **Identifying adversaries:** Organizations must be aware of the various threat actors that could target them, including nation-state actors, cybercriminals, hacktivists, and insider threats. By understanding who their potential adversaries are, organizations can better anticipate and prepare for attacks.

2. **Understanding motives and objectives:** Different threat actors have varying motives and objectives, ranging from financial gain to political activism, espionage, or sabotage. By grasping the driving forces behind their adversaries' actions, organizations can predict potential targets and develop effective defense strategies.

3. **Studying tactics, techniques, and procedures (TTPs):** Cyber attackers employ a wide range of TTPs to infiltrate networks, compromise systems, and exfiltrate data. By

analyzing and understanding these TTPs, organizations can identify patterns, recognize potential vulnerabilities, and implement appropriate countermeasures.

4. **Leveraging threat intelligence:** To stay informed about the latest threats, organizations should make use of threat intelligence sources, such as threat feeds, vulnerability databases, and information sharing platforms. This enables them to stay updated on emerging threats, vulnerabilities, and trends in the cybersecurity landscape.

Knowing Yourself

In addition to understanding the enemy, organizations must also possess a thorough knowledge of their own strengths and weaknesses. This self-awareness is crucial for effective cybersecurity, as it enables organizations to optimize their security posture and allocate resources effectively. Key aspects of knowing yourself include:

1. **Assessing internal assets:** Organizations need to have a clear understanding of their critical assets, including hardware, software, data, and intellectual property. By identifying and prioritizing these assets, organizations can allocate resources and implement security measures accordingly.

2. **Evaluating vulnerabilities:** Organizations must be aware of their own vulnerabilities, including technical, human, and process-related weaknesses. By conducting regular vulnerability assessments and penetration tests, organizations can identify and remediate these weaknesses to reduce their attack surface.

3. **Understanding internal processes and policies:** A clear understanding of an organization's security policies, processes, and procedures is essential to ensure that they are effective and up to date. Regular reviews and updates of these policies and processes can help organizations maintain a strong cybersecurity posture.

4. **Building a strong cybersecurity culture:** One of the most significant aspects of knowing yourself is fostering a cybersecurity-aware culture within the organization. This

involves training and educating employees, promoting security best practices, and encouraging a proactive approach to cybersecurity.

The principle of knowing your enemy and yourself is an essential component of the OODA Loop and effective cybersecurity. By cultivating a deep understanding of both adversaries and their own organization, security professionals can develop robust strategies and tactics to defend against cyber threats and maintain a strong security posture.

Deception and Indirect Approaches: The Art of Subterfuge in the OODA Loop and Cybersecurity

Deception and indirect approaches have long been hallmarks of warfare and conflict, as evidenced in Sun Tzu's "The Art of War." In the context of cybersecurity, these tactics can also prove valuable for both offense and defense. By incorporating deception and indirect approaches into the OODA Loop, organizations can enhance their cybersecurity posture, confuse adversaries, and protect critical assets more effectively.

Deception in Cybersecurity

Deception is the art of deliberately misleading adversaries to gain a tactical or strategic advantage. In the realm of cybersecurity, deception can take various forms, including:

- **Honeypots and honeynets:** Honeypots are decoy systems designed to mimic real assets and attract attackers. Honeynets are networks of honeypots designed to appear as real targets. Both honeypots and honeynets can be used to gather intelligence on attackers, study their tactics, techniques, and procedures (TTPs), and buy time to protect real assets.

- **Misdirection:** Misdirection techniques involve creating false trails and breadcrumbs to lead attackers away from actual targets. This can include fake network shares, bogus user accounts, or misleading file names that distract attackers and waste their resources.

- **Disinformation:** By injecting false or misleading information into the cyber environment, organizations can deceive adversaries and potentially cause them to make strategic errors. Disinformation can include fake communications, deceptive social media profiles, or falsified threat intelligence.

Indirect Approaches in Cybersecurity

Indirect approaches, in contrast to deception, involve avoiding head-on confrontations and instead exploiting an adversary's weaknesses or vulnerabilities. In cybersecurity, indirect approaches can include:

- **Exploiting gaps in an attacker's knowledge:** By understanding an adversary's blind spots and limitations, organizations can develop strategies to exploit these gaps. This may involve using uncommon technologies, protocols, or configurations that the attacker is less familiar with or creating complex network topologies that are difficult to navigate.
- **Leveraging an attacker's strengths against them:** Cyber attackers often have specific TTPs that they rely on due to their success in previous campaigns. By identifying these strengths and turning them against the attacker, organizations can catch them off guard and undermine their attack.
- **Focusing on asymmetric advantages:** Rather than directly confronting an attacker, organizations can leverage their unique advantages, such as superior threat intelligence, a strong security culture, or innovative defensive technologies, to create an asymmetric advantage that makes it more difficult for the attacker to succeed.

Incorporating Deception and Indirect Approaches into the OODA Loop

The OODA Loop, with its emphasis on continuous observation, orientation, decision-making, and action, provides an ideal framework for incorporating deception and indirect approaches into an organization's cybersecurity strategy. By integrating these tactics into each Phase of the loop, organizations can enhance their ability to anticipate and respond to cyber threats.

For example, during the Observe Phase, organizations can use honeypots and honeynets to gather valuable intelligence on potential attackers. In the Orient Phase, this intelligence can be used to identify gaps in the attacker's knowledge or capabilities, which can then be exploited in the Decide and Act Phases through misdirection, disinformation, or other indirect tactics.

Deception and indirect approaches offer valuable tools for organizations looking to enhance their cybersecurity posture and better defend against cyber threats. By incorporating these tactics into the OODA Loop, organizations can create a dynamic and adaptable cybersecurity strategy that keeps adversaries guessing and maintains a strong security posture.

Strategic Positioning: Leveraging the OODA Loop for Optimal Cybersecurity Posture

Strategic positioning, a concept with roots in military strategy and "The Art of War," is the process of positioning one's resources and capabilities to create a favorable environment for success. In the context of cybersecurity, strategic positioning involves the deliberate alignment of an organization's security infrastructure, policies, and practices to optimize its defense against cyber threats. By incorporating the principles of the OODA Loop into strategic positioning, organizations can create a dynamic and resilient cybersecurity posture that is adaptable to evolving threats and environments.

Understanding the Cyber Threat Landscape

The first step in strategic positioning is to develop a comprehensive understanding of the cyber threat landscape. This involves continuous observation and monitoring of the internal and external environments to identify emerging threats, vulnerabilities, and trends. Organizations should gather and analyze threat intelligence from a variety of sources, such as open-source reports, industry partners, and government agencies, to ensure a well-rounded understanding of the threats that apply to their unique environment.

Adapting Security Posture Based on Threat Landscape

Once an organization has developed a clear understanding of the cyber threat landscape, it can begin to orient its security posture to address the most significant risks. This involves aligning security policies, processes, and technologies to effectively mitigate these risks. For example, an organization that identifies an increase in spear-phishing attacks targeting its employees may choose to prioritize employee security awareness training and implement stronger email security controls.

The OODA Loop's emphasis on continuous decision-making and action is critical to ensuring that an organization's security posture remains adaptive and responsive to changing threats. By regularly evaluating and adjusting their security posture based on the current threat landscape, organizations can maintain a strong defense against cyber threats.

Leveraging Unique Advantages and Capabilities

Strategic positioning also involves identifying and leveraging an organization's unique advantages and capabilities to create a competitive edge in cybersecurity. This may include technological innovations, a strong security culture, or access to specialized expertise. By capitalizing on these strengths, organizations can build a more robust and resilient security posture that is difficult for adversaries to overcome.

For example, an organization with a highly skilled incident response team may choose to invest in advanced threat hunting capabilities to proactively identify and mitigate potential security incidents. In contrast, an organization with a strong security culture may prioritize employee security awareness training and initiatives to foster a security-conscious workforce.

Collaboration and Information Sharing

Another important aspect of strategic positioning in cybersecurity is fostering collaboration and information sharing among internal teams, industry partners, and government agencies. By establishing strong relationships and communication channels, organizations can share threat intelligence, best practices, and lessons learned to enhance their collective security posture. This collaborative approach helps create a more comprehensive understanding of the cyber threat landscape and enables organizations to better anticipate and respond to potential threats.

Strategic positioning is a critical component of an organization's cybersecurity strategy, and the OODA Loop provides an ideal framework for incorporating it. By continuously observing the threat landscape, orienting security posture based on identified risks, leveraging unique advantages, and fostering collaboration and information sharing, organizations can create a dynamic and resilient cybersecurity posture that is well-equipped to handle the ever-evolving cyber threat landscape.

Exploiting Opportunities: Harnessing the OODA Loop for Proactive Cybersecurity

This discussion shows how cybersecurity professionals and leaders can leverage the key concepts considered throughout this book to proactively assess and manage risks.

In the rapidly evolving world of cybersecurity, organizations must be adept at identifying and exploiting opportunities to enhance their security posture and stay ahead of potential threats. By integrating the principles of the OODA Loop into their cybersecurity strategies, organizations can proactively seek out and capitalize on opportunities to improve their defenses, mitigate risks, and ultimately outmaneuver adversaries in cyberspace.

Identifying Opportunities for Improvement

One of the key aspects of exploiting opportunities in cybersecurity is the ability to identify areas where improvements can be made. This requires a constant process of observation and analysis, as well as an openness to change and innovation. Organizations should regularly assess their security posture, evaluate the effectiveness of their policies and technologies, and consider potential areas for growth or refinement.

Through the OODA Loop's iterative process, organizations can continuously identify and assess opportunities to strengthen their cybersecurity defenses. This might involve adopting innovative technologies or techniques, refining existing processes, or implementing novel approaches to proactively address emerging threats.

Innovation and Experimentation

A critical component of exploiting opportunities in cybersecurity is fostering a culture of innovation and experimentation. This involves encouraging and supporting the development of latest ideas, technologies, and strategies to address emerging threats and vulnerabilities. By embracing a mindset of continuous improvement and adaptation, organizations can stay ahead of the curve and maintain a proactive cybersecurity posture.

Organizations can also experiment with different security tools, strategies, and configurations to determine which approaches are most effective in their specific environments. By conducting regular tests and evaluations, security teams can identify potential gaps or weaknesses in their defenses and take corrective action to address them.

Leveraging Threat Intelligence and Emerging Technologies

To exploit opportunities in cybersecurity, organizations must stay informed about the latest threat intelligence, emerging technologies, and industry trends. This includes gathering and analyzing information on new attack vectors, vulnerabilities, and threat actors, as well as staying abreast of advancements in security tools and technologies.

By incorporating this knowledge into the OODA Loop process, organizations can better anticipate and respond to emerging threats, as well as identify opportunities to enhance their defenses. For example, an organization that becomes aware of a new malware variant targeting their industry may choose to invest in advanced threat detection and response capabilities to protect against similar threats in the future.

Collaboration and Information Sharing

Exploiting opportunities in cybersecurity is a critical aspect of maintaining a proactive and resilient security posture. This effort often requires collaboration and information sharing among internal teams, industry partners, and government agencies. By establishing strong relationships and communication channels, organizations can share threat intelligence, best practices, and lessons learned to enhance their collective security posture. This collaborative approach helps create a more comprehensive understanding of the cyber threat landscape and enables organizations to better anticipate and respond to potential threats.

By integrating the OODA Loop into their strategies, organizations can continuously identify, assess, and capitalize on opportunities to strengthen their defenses and stay ahead of potential threats in the constantly evolving world of cyberspace.

Unity of Effort and Command: Strengthening Cybersecurity through the OODA Loop

In the complex and rapidly changing landscape of cybersecurity, organizations must maintain a unified approach to address emerging threats and vulnerabilities effectively. Unity of effort and command are also essential principles for ensuring a coherent, coordinated, and efficient response to cyber incidents. By incorporating these concepts into implementing the OODA Loop framework, organizations can enhance their cybersecurity posture and improve their ability to detect, prevent, and respond to cyber threats.

Establishing Clear Leadership and Decision-Making Structures

One of the key aspects of unity of effort and command in cybersecurity is having a clear leadership structure and well-defined decision-making processes. This helps to ensure that all members of the organization understand their roles and responsibilities and can collaborate effectively in addressing cyber threats.

By aligning the OODA Loop with organizational hierarchies and decision-making structures, organizations can streamline the flow of information, improve situational awareness, and facilitate rapid decision-making in response to cyber incidents. This enables security teams to act more swiftly and decisively in identifying, analyzing, and mitigating threats.

Fostering Collaboration and Coordination across Teams

Unity of effort and command also involves fostering collaboration and coordination across different teams within an organization. This is particularly important in the realm of cybersecurity, where threats can impact multiple areas of the business and require input from various stakeholders.

By integrating the OODA Loop into interdepartmental processes and communication channels, organizations can improve collaboration and information sharing among security teams, IT teams, and other business units. This helps to create a more holistic understanding of the threat landscape and enables organizations to coordinate their efforts more proactively to address cyber risks more effectively.

Aligning Security Strategies and Objectives

By incorporating the OODA Loop into their strategic planning processes, organizations can continuously assess their cybersecurity posture and make informed decisions about how to allocate resources and prioritize initiatives. To maintain unity of effort and command in cybersecurity, organizations must ensure that their security strategies and objectives are aligned with their overall business goals and priorities. This involves regularly reviewing and updating security policies, processes, and technologies to ensure they remain effective in the face of evolving threats and business requirements. This helps to ensure that all cybersecurity efforts are aligned with the organization's broader objectives and contribute to a unified approach to risk management.

Standardizing Processes and Procedures

To achieve unity of effort and command, organizations must also standardize their processes and procedures for addressing cybersecurity risks. This includes establishing clear guidelines, protocols, and workflows for tasks such as threat detection, incident response, and vulnerability management. By embedding the OODA Loop into these standardized processes and procedures, organizations can improve the consistency and efficiency of their cybersecurity efforts. This not only helps to ensure that all team members are working in a coordinated manner but also enables organizations to measure and track the success of their security initiatives more effectively.

Unity of effort and command are essential principles for strengthening cybersecurity in today's complex and rapidly evolving threat landscape. By incorporating these concepts into the OODA Loop framework, organizations can improve their ability to detect, prevent, and respond to cyber threats in a more coordinated, efficient, and effective manner.

What about The Book of 5 Rings?

Miyamoto Musashi's "The Book of Five Rings" (Go Rin No Sho) is a renowned classic treatise on martial arts and strategy, written in the 17th century by the famous Japanese swordsman Miyamoto Musashi. The book is organized into five sections or "rings," each representing an element of battle and life: Earth, Water, Fire, Wind, and Void. While it primarily focuses on martial arts, the principles discussed in the book have been applied to various fields such as business, politics, and personal development.

1. **Earth:** In this section, Musashi lays the foundation of his teachings by emphasizing the importance of understanding the principles of martial arts and the development of a strategic mindset. He advocates learning from different martial arts schools, understanding the strengths and weaknesses of each, and adapting them to one's own style. He stresses the importance of practicing and refining techniques, cultivating a strong spirit, and developing a deep understanding of the underlying principles of combat.

2. **Water:** This section focuses on the flexibility and adaptability of the warrior, comparing their mindset to the flowing nature of water. Musashi highlights the importance of being fluid in one's movements and adapting to the opponent's actions. He encourages the warrior to maintain a calm and composed mindset, even in the heat of battle, allowing them to make rational decisions and execute strategies effectively.

3. **Fire:** Here, Musashi discusses the importance of timing and tempo in combat. He emphasizes the need to control the

173

pace of a confrontation, exploiting the opponent's weaknesses and mistakes, and striking at the most opportune moments. The warrior must maintain their composure and avoid becoming overly aggressive or reckless.

4. **Wind:** In this section, Musashi examines the various martial arts schools and styles, comparing their techniques and principles. He encourages the warrior to study and understand the strengths and weaknesses of different styles, incorporating the most effective techniques into their own arsenal. By doing so, the warrior can adapt to any situation and opponent.

5. **Void:** The final section discusses the concept of the "Void" or "No-thingness," referring to a state of mind where the warrior transcends conscious thought and achieves a heightened level of awareness and intuition. In this state, the warrior can perceive the subtlest details of their environment and the movements of their opponent, allowing them to react swiftly and effectively.

Comparisons to the OODA Loop:

While Musashi's "The Book of Five Rings" and John Boyd's OODA Loop originate from diverse cultural backgrounds and focus on different subjects, several similarities can be drawn between the two:

- **Adaptability:** Both Musashi's teachings and the OODA Loop emphasize the importance of adaptability, flexibility, and the ability to respond effectively to changing situations. In the context of cybersecurity, this means continually updating strategies and tactics in response to evolving threats.

- **Understanding the enemy:** Musashi and Boyd both stress the importance of knowing one's opponent to achieve success. In cybersecurity, this translates to gathering threat intelligence, understanding the tactics and techniques of adversaries, and anticipating their next moves.

- **Timing:** Musashi's focus on timing and tempo in combat can be compared to the OODA Loop's emphasis on speed and agility in decision-making processes. In cybersecurity, the ability to detect, analyze, and respond to threats quickly can be the difference between a successful defense and a damaging security breach.
- **Continuous improvement:** Both Musashi and the OODA Loop advocate for constant learning, refinement, and improvement of one's techniques and strategies. This is especially relevant in the rapidly changing field of cybersecurity, where new threats and vulnerabilities are constantly emerging.

By incorporating Musashi's emphasis on adaptability, understanding the enemy, timing, and continuous improvement, cybersecurity professionals can enhance their application of the OODA Loop, enabling them to better protect their organizations from ever-evolving cyber threats.

Both the OODA Loop and the principles discussed in "The Book of Five Rings" highlight the importance of situational awareness and information gathering. In the context of cybersecurity, this means continuously monitoring and collecting data about the threat landscape and sharing that information within the organization and with relevant stakeholders. Effective communication and collaboration play a crucial role in fostering a collective understanding of the threats and vulnerabilities faced by an organization, enhancing its overall cybersecurity posture.

Furthermore, Musashi's concept of the "Void" shares similarities with the OODA Loop's emphasis on situational awareness and rapid decision-making. Achieving a heightened level of awareness and intuition allows cybersecurity professionals to perceive subtle signs of potential threats and react swiftly to mitigate risks. This state of mind is particularly important in today's complex and fast-paced digital environment, where new threats can emerge and evolve rapidly.

By embracing the principles outlined in Miyamoto Musashi's "The Book of Five Rings" and applying them to the OODA Loop framework, cybersecurity professionals can develop a more comprehensive and effective approach to managing cybersecurity risks. This fusion of ancient wisdom and modern strategic thinking can serve as a powerful tool for navigating the constantly shifting landscape of cybersecurity threats and vulnerabilities, ensuring that organizations remain resilient and secure in the face of an ever-changing digital world.

More Sources

1. Boyd, J. R. (1996). The Essence of Winning and Losing. Retrieved from **https://dnipogo.org/john-r-boyd/**
2. Boyd, J. R. (1976). Destruction and Creation. Retrieved from **https://upload.wikimedia.org/wikipedia/commons/a/a6/Destruction_%26_Creation.pdf**
3. **Hammond, G. T. (2001). The Mind of War: John Boyd and American Security. Smithsonian Institution Press.**
4. **Richards, C. (2004). Certain to Win: The Strategy of John Boyd Applied to Business. Xlibris Corporation.**
5. The Art of Manliness. (2019). The OODA Loop: How to Turn Uncertainty into Opportunity. Retrieved from **https://www.artofmanliness.com/character/behavior/ooda-loop/**
6. **Osinga, F. P. (2005). Science, Strategy, and War: The Strategic Theory of John Boyd. Routledge.**
7. Spinosa, E. (2020). Boyd's OODA Loop (It's Not What You Think). Retrieved from **https://fasttransients.files.wordpress.com/2020/03/boydsoodaloopnecesse-1.pdf**
8. Case studies:
   1. **Stuxnet: Langner, R. (2011). Stuxnet: Dissecting a Cyberwarfare Weapon. IEEE Security & Privacy, 9(3), 49-51. doi:10.1109/MSP.2011.67**
   2. Sony Pictures Hack: Perlroth, N., & Sanger, D. E. (2014). North Korea Loses Its Link to the Internet. The New York Times. Retrieved from **https://www.nytimes.com/2014/12/23/world/asia/attack-is-suspected-as-north-korean-internet-collapses.html**

3. WannaCry Ransomware: NIST. (2018). National Institute of Standards and Technology. Retrieved from **https://www.nccoe.nist.gov/sites/default/files/library/sp1800/cr-recovery-guide-sp1800-11.pdf**

9. Cyber Strategist. (2009-2010). Five Keys to Successful Cyber Security Strategy. Retrieved from **https://cyberstrategist.wordpress.com**

10. Sun Tzu. (n.d.). The Art of War. Retrieved from **https://suntzusaid.com/download.php**

Epilogue: Unleashing the Power of the OODA Loop in Cybersecurity

"Effective cybersecurity requires not just an understanding of technology, but an understanding of human nature." - Bruce Schneier

In the ever-evolving landscape of cybersecurity, organizations are under constant pressure to adapt and respond to an array of sophisticated cyber threats. As these threats become increasingly diverse and complex, traditional security measures often struggle to keep pace with the rapid changes. It is in this challenging environment that the OODA Loop, a powerful decision-making model originally developed by military strategist John Boyd, emerges as a transformative tool for cybersecurity professionals. This compendium of knowledge, insights, and best practices aims to provide an extensive and comprehensive exploration of the OODA Loop's immense potential in revolutionizing the field of cybersecurity.

The value of the OODA Loop in cybersecurity cannot be overstated. Its ability to enhance situational awareness, streamline decision-making processes, and enable swift, adaptive responses to emerging threats makes it a critical asset in the constant battle against cyber adversaries. Through the meticulous examination of various concepts, strategies, and case studies, this comprehensive resource delves deep into the OODA Loop's core principles and illustrates their practical applications within the realm of cybersecurity.

The OODA Loop, comprising four Phases—Observe, Orient, Decide, and Act—is a powerful, iterative decision-making model that emphasizes speed, adaptability, and continuous learning. The higher-level topics explored within this work include the importance of observation and situational awareness, the role of orientation in shaping perceptions and informing decisions, the decision-making process itself, and the execution of chosen actions. Throughout these discussions, recurring themes and vital principles emerge, underlining their importance in successfully applying the OODA Loop to cybersecurity.

One of these recurring themes is the significance of cultural traditions, genetic heritage, new information, and previous experiences in the application of the OODA Loop to cybersecurity. These factors play a crucial role in shaping an individual's understanding of their environment and their ability to adapt to new and emerging threats. By acknowledging and accounting for these influences, cybersecurity professionals can enhance their ability to think critically and creatively, enabling them to develop more effective strategies and solutions.

Another critical aspect of the OODA Loop is the concept of Implicit Guidance and Control. This principle emphasizes the importance of intuition, experience, and trust in facilitating rapid decision-making and action. By fostering an environment that encourages the development of these skills and the cultivation of a deep understanding of the cyber threat landscape, organizations can significantly accelerate their OODA Loop processes, leading to improved incident resolution and a more robust cybersecurity posture.

- **Feedback and feedforward mechanisms** within the OODA Loop are also essential components in the successful application of this model to cybersecurity. These mechanisms enable continuous learning and adaptation, ensuring that security teams can refine their strategies and tactics based on the outcomes of previous actions and the evolving threat landscape. By harnessing the power of feedback and feedforward, organizations can create a more resilient and adaptable cybersecurity program capable of withstanding even the most sophisticated and persistent cyber-attacks.

- **Integration with other cybersecurity frameworks and methodologies**, such as the NIST Cybersecurity Framework, MITRE ATT&CK, and the ISO/IEC 27000 series of standards, is another vital aspect of applying the OODA Loop to cybersecurity. By incorporating the OODA Loop into existing frameworks and methodologies, organizations can create a more comprehensive approach to cybersecurity risk management, ensuring that all aspects of their security posture are effectively addressed and continuously improved.

- **Threat intelligence plays a critical role in the OODA Loop**, particularly in the Observe and Orient Phases. By leveraging timely, accurate, and actionable information about emerging threats and vulnerabilities, organizations can enhance their situational awareness and improve their ability to make informed decisions about how to address potential risks. This integration of threat intelligence into the OODA Loop process can significantly improve an organization's cybersecurity capabilities and resilience against cyber-attacks.

- **Automation and artificial intelligence (AI)** have the potential to revolutionize the OODA Loop process, particularly in the Observe, Orient, and Act Phases. The rapid advancements in these technologies can enhance tasks like threat detection, analysis, and remediation, allowing organizations to accelerate their OODA Loop and respond more effectively to cyber threats. By embracing automation and AI, cybersecurity professionals can unlock new levels of efficiency and effectiveness in their ongoing battle against cyber adversaries

- **The OODA Loop's application in distributed environments**, such as remote work, cloud computing, and distributed systems, is another important topic in today's increasingly interconnected world. By adapting the OODA Loop to these environments, organizations can effectively manage cybersecurity risks and protect their valuable digital assets, regardless of where they reside.

Measuring the effectiveness of the OODA Loop within a cybersecurity program is essential to ensure its continued success. By identifying key performance indicators (KPIs) and metrics, organizations can track and improve their OODA Loop implementation, fine-tuning their processes to maximize their cybersecurity capabilities.

Human factors, such as cognitive biases, decision-making styles, and situational awareness, are integral to the successful application of the OODA Loop in cybersecurity. By acknowledging and addressing these factors, organizations can enhance the effectiveness of their security teams and ensure that they are better equipped to respond to emerging threats. Legal and ethical considerations must be considered when implementing the OODA Loop in cybersecurity. Compliance with data protection laws and regulations, such as GDPR and CCPA, is crucial, as is ensuring that security actions align with ethical guidelines and do not infringe on the privacy rights of individuals or the intellectual property of other organizations.

- **The principles of resilience and adaptability are central to the OODA Loop's effectiveness in cybersecurity.** Organizations must be prepared to adapt their strategies, tactics, and tools as new threats emerge and the threat landscape changes. This may involve regularly reviewing and updating security policies, processes, and technologies to ensure they remain effective in the face of new challenges.

Effective communication and collaboration are vital to the successful application of the OODA Loop in cybersecurity. Security teams should establish clear channels for sharing information and insights, both internally and externally, to enhance the collective understanding of the threat landscape and improve the organization's overall cybersecurity posture.

- **Lastly, post-incident analysis and learning are essential components of the OODA Loop process.** After a security incident has been resolved, organizations must conduct thorough post-incident analyses to identify lessons learned and areas for improvement. This process can help refine the OODA Loop and ensure that security teams are better prepared to respond to future threats.

In conclusion, this extensive exploration of the OODA Loop's potential in cybersecurity reveals a wealth of knowledge, insights, and best practices that can be harnessed by security professionals to revolutionize their approach to defending against cyber threats. By understanding and embracing the power of the OODA Loop, organizations can unlock new levels of efficiency, adaptability, and resilience in the face of an ever-changing cyber landscape. This work serves as a testament to the immense value of the OODA Loop in cybersecurity and a guide for those seeking to leverage its transformative potential in their organizations.

APPENDIX A: Check List

A comprehensive approach to using the OODA Loop for cybersecurity.

1. **Establish a strong cybersecurity team:** Assemble a team of skilled professionals with expertise in various aspects of cybersecurity to monitor, assess, and respond to threats.
2. **Develop an effective threat intelligence program:** Collect and analyze data from various sources to identify, track, and understand emerging threats and vulnerabilities.
3. **Implement continuous monitoring and detection:** Employ monitoring and detection tools, such as intrusion detection systems and security information and event management (SIEM) systems, to identify potential cyber threats and incidents.
4. **Foster a culture of collaboration and communication:** Encourage information sharing and collaboration both internally and externally, promoting a sense of shared responsibility for cybersecurity.
5. **Continuously assess and update security controls:** Regularly review and adjust security controls to maintain an optimal balance between security and usability, considering the evolving threat landscape.
6. **Prioritize actions based on risk:** Allocate resources and focus efforts on the most critical threats and vulnerabilities, considering their likelihood and potential impact on the organization.
7. **Develop and maintain incident response plans:** Establish clear procedures and guidelines for detecting, containing, eradicating, and recovering from security incidents, ensuring timely and effective action.
8. **Train and educate employees:** Provide ongoing training and awareness programs to ensure that employees understand their role in maintaining the organization's cybersecurity posture and can recognize and respond to potential threats.

9. **Adopt an agile approach to cybersecurity:** Embrace flexibility, adaptability, and continuous improvement, regularly reviewing and updating security policies, procedures, and technologies.
10. **Leverage automation and advanced technologies:** Utilize automated tools, artificial intelligence, and advanced analytics to enhance threat detection, decision-making, and incident response capabilities.
11. **Align with legal and regulatory requirements:** Ensure that cybersecurity strategies and actions comply with relevant laws and regulations, conducting regular compliance audits and working closely with legal counsel.
12. **Conduct regular reviews and debriefs:** Perform post-incident reviews to identify lessons learned, update security policies and procedures, and continuously improve the organization's cybersecurity capabilities.

And just for good measure, ChatGPT added:

13. **Establish backup and recovery plans:** Develop and maintain strategies for data backup and system recovery to minimize downtime and data loss in the event of a cybersecurity incident.
14. **Conduct penetration testing and vulnerability assessments:** Regularly test your organization's security measures and infrastructure to identify and address weaknesses before they can be exploited.
15. **Engage with industry partners and cybersecurity communities:** Stay informed about the latest trends, best practices, and threat intelligence by participating in industry events, forums, and professional organizations.
16. **Implement a robust access control policy:** Ensure that access to critical systems and sensitive information is limited to authorized personnel and use multi-factor authentication to enhance security.
17. **Keep software and hardware up to date:** Regularly update all software, operating systems, and hardware components to ensure they are protected against known vulnerabilities.

18. **Develop a comprehensive cybersecurity policy:** Create and maintain a formal cybersecurity policy that outlines the organization's goals, strategies, and responsibilities, as well as the acceptable use of technology.

REFERENCE: Question: "Please review my checklist and add any additional references you can think of?" Date:2023-03-12 Site: HTTPS://chat.openai.com/

APPENDIX B: Other Books and Articles by the Author

Closing the Security Agility Gap. IEEE · Aug 19, 2022
What Every Engineer Should Know About Cyber Security and Digital Forensics 2nd Edition
Third Party Risk Management (TPRM) – A Series in Program Development - Part 1
Part 2: Third Party Risk Management (TPRM) – A Series in Program Development
Part 3: Third Party Risk Management (TPRM) – A Series in Program Development
Black Kite Blogs - https://blackkite.com/?s=Maley&post_type=post

APPENDIX C: OODA Loop Reading List

Some of the most well-known and comprehensive books on the OODA Loop and its applications in various fields, including military strategy, business, and decision-making.

1. "Science, Strategy and War: The Strategic Theory of John Boyd" by Frans P.B. Osinga
2. "Boyd: The Fighter Pilot Who Changed the Art of War" by Robert Coram
3. "The OODA Loop: A Comprehensive Guide to the OODA Loop and Decision Making" by Christopher P. Nemeth
4. "Certain to Win: The Strategy of John Boyd, Applied to Business" by Chet Richards
5. "The Tao of Boyd: How to Master the OODA Loop" by Harry Hillaker
6. "The Quick and Dirty Guide to Boyd's OODA Loop" by David Malone
7. "The Art of Maneuver: Maneuver-Warfare Theory and Airland Battle" by Robert R. Leonhard
8. "Tempo: Timing, Tactics, and Strategy in Narrative-driven Decision-making" by Venkatesh G. Rao
9. "The Mind of War: John Boyd and American Security" by Grant Tedrick Hammond
10. "The OODA Loop for Intelligence Analysis: A Guidebook for Analysts and Decision Makers" by Michael G. Harris

Here are additional books that discuss the OODA Loop or incorporate the concept in their content, focusing on various fields such as business, military strategy, and decision-making. Please note that while these books may not be solely focused on the OODA Loop, they contain significant content, analysis, or discussions related to the concept and its applications in various fields.

11. "Adaptive Thinking: How to Apply the OODA Loop to Your Everyday Life" by Jack D. Kem
12. "Revelations of the OODA Loop" by Shawn Spears

13. "Agile Software Development with Scrum: A Practical Guide" by Ken Schwaber and Mike Beedle - This book discusses Scrum, an agile software development methodology that incorporates the OODA Loop principles.
14. "Left of Bang: How the Marine Corps' Combat Hunter Program Can Save Your Life" by Patrick Van Horne and Jason A. Riley - This book covers the Combat Hunter Program, which uses the OODA Loop to enhance situational awareness and decision-making in dangerous environments.
15. "OODA Loop: A Comprehensive Guide to the OODA Loop and Decision-Making" by Paul Russell
16. "OODA Loop: How to Improve Decision-Making Speed and Accuracy" by Jack Morgan
17. "Fast Transients: Lessons in Strategy and Decision Making from the World of Motorsports" by Steve Matchett - This book discusses the application of the OODA Loop in motorsports and its relevance to strategy and decision-making.
18. "OODA: The Art of Agile Decision Making" by Michael Simmons
19. "Winning in FastTime: Harness the Competitive Advantage of Prometheus in Business and Life" by John A. Warden III and Leland Russell - This book discusses the application of the OODA Loop in business strategy and decision-making.
20. "Leading at the Edge: Leadership Lessons from the Extraordinary Saga of Shackleton's Antarctic Expedition" by Dennis N. T. Perkins - This book explores leadership lessons that can be derived from Shackleton's expedition, incorporating the OODA Loop concept in decision-making.

APPENDIX D: Information Sharing and Analysis Centers

Information Sharing and Analysis Centers (ISACs) are industry-specific organizations that facilitate collaboration and information sharing on cybersecurity threats, vulnerabilities, and best practices among their members, which typically include companies and organizations within a specific sector. ISACs work to improve the overall security posture of their respective sectors by fostering communication and cooperation among members, as well as between the public and private sectors. They collect, analyze, and disseminate actionable intelligence, enabling members to enhance their cyber defense capabilities, protect critical infrastructure, and better respond to emerging threats. Additionally, ISACs often collaborate with government agencies, such as the Department of Homeland Security (DHS), to provide a more comprehensive understanding of the cyber threat landscape.

The following is a list of many ISACs.

- **Health Information Sharing and Analysis Center (H-ISAC)**
- **Information Sharing and Analysis Center (ISAC)**
- **Information Sharing and Analysis Organizations (ISAOs)**
- **Multi-State Information Sharing and Analysis Center (MS-ISAC)**
- **Aviation-ISAC (A-ISAC)**
- **Auto-ISAC**
- **Communications-ISAC (Comm-ISAC)**
- **Defense Industrial Base-ISAC (DIB-ISAC)**
- **Downstream Natural Gas-ISAC (DNG-ISAC)**
- **Electricity-ISAC (E-ISAC)**
- **Financial Services-ISAC (FS-ISAC)**
- **Information Technology-ISAC (IT-ISAC)**
- **Legal Services-ISAC (LS-ISAC)**
- **Maritime-ISAC (M-ISAC)**
- **National Council of ISACs (NCI)**
- **Oil and Natural Gas-ISAC (ONG-ISAC)**
- **Real Estate-ISAC (RE-ISAC)**
- **Research and Education Network-ISAC (REN-ISAC)**
- **Retail & Hospitatility ISAC - (RH-ISAC)**
- **Surface Transportation-ISAC (ST-ISAC)**
- **Water-ISAC (W-ISAC)**

Keep in mind that the landscape of ISACs can change over time. To find the most current information, it's recommended that you consult official sources or visit the National Council of ISACs **website** for any updates.

APPENDIX F: Bonus Material – Exam

You can find the exam and answer key at **https://www.c-ooda.com/exam**

Made in the USA
Las Vegas, NV
03 April 2024

88191151R00105